Leveled Texts for Science

Physical Science

SHELL EDUCATION

Reading Level Editor
Josh BishopRoby

English Language Learner Consultants
D. Kyle Shuler
Chino Valley Unified School District, California
Marcela von Vacano
Arlington County Schools, Virginia

Gifted Education Consultant
Wendy Conklin, M.A.
Mentis Online, Round Rock, Texas

Special Education Consultant
Dennis Benjamin
Prince William County Public Schools, Virginia

Contributing Content Authors
Gina dal Fuoco
Connie Jankowski
Lisa E. Greathouse
Greg Young
Lynn Van Gorp, M.S.
William B. Rice
Debra J. Housel

Publisher
Corinne Burton, M.A.Ed.

Editorial Director
Dona Herweck Rice

Creative Director
Lee Aucoin

Editor-in-Chief
Sharon Coan, M.S.Ed.

Editorial Manager
Gisela Lee, M.A.

Cover Art
Lesley Palmer

Print Production
Neri Garcia

Shell Education
5301 Oceanus Drive
Huntington Beach, CA 92649
http://www.shelleducation.com
ISBN 978-1-4258-0161-8

Reprinted 2010

Table of Contents

What Is Differentiation?

Over the past few years, classrooms have evolved into diverse pools of learners. Gifted students, English language learners, special needs students, high achievers, underachievers, and average students all come together to learn from one teacher. The teacher is expected to meet their diverse needs in one classroom. It brings back memories of the one-room schoolhouse during early American history. Not too long ago, lessons were designed to be one size fits all. It was thought that students in the same grade level learned in similar ways. Today, we know that viewpoint to be faulty. Students have differing learning styles, come from different cultures, experience a variety of emotions, and have varied interests. For each subject, they also differ in academic readiness. At times, the challenges teachers face can be overwhelming, as they struggle to figure out how to create learning environments that address the differences they find in their students.

What is differentiation? Carol Ann Tomlinson at the University of Virginia says, "Differentiation is simply a teacher attending to the learning needs of a particular student or small group of students, rather than teaching a class as though all individuals in it were basically alike" (2000). Differentiation can be carried out by any teacher who keeps the learners at the forefront of his or her instruction. The effective teacher asks, "What am I going to do to shape instruction to meet the needs of all my learners?" One method or methodology will not reach all students.

Differentiation encompasses what is taught, how it is taught, and the products students create to show what they have learned. When differentiating curriculum, teachers become the organizers of learning opportunities within the classroom environment. These categories are often referred to as content, process, and product.

- **Content:** Differentiating the content means to put more depth into the curriculum through organizing the curriculum concepts and structure of knowledge.
- **Process:** Differentiating the process requires the use of varied instructional techniques and materials to enhance the learning of students.
- **Product:** When products are differentiated, cognitive development and the students' abilities to express themselves improves.

Teachers should differentiate content, process, and product according to students' characteristics. These characteristics include students' readiness, learning styles, and interests.

- **Readiness:** If a learning experience aligns closely with students' previous skills and understanding of a topic, they will learn better.
- **Learning styles:** Teachers should create assignments that allow students to complete work according to their personal preferences and styles.
- **Interests:** If a topic sparks excitement in the learners, then students will become involved in learning and better remember what is taught.

How to Differentiate Using This Product

The leveled texts in this series help teachers differentiate science content for their students. Each book has 15 topics, and each topic has a text written at four different reading levels. (See page 17 for more information.) These texts are written at a variety of reading levels, but all the levels remain strong in presenting the science content and vocabulary. Teachers can focus on the same content standard or objective for the whole class, but individual students can access the content at their instructional levels rather than at their frustration levels.

Determining your students' instructional reading levels is the first step in the process. It is important to assess their reading abilities often so they do not get tracked into one level. Below are suggested ways to use this resource, as well as other resources in your building, to determine students' reading levels.

- **Running records:** While your class is doing independent work, pull your below-grade-level students aside, one at a time. Have them read aloud the lowest level of a text (the star level) individually as you record any errors they make on your own copy of the text. If students read accurately and fluently and comprehend the material, move them up to the next level and repeat the process. Following the reading, ask comprehension questions to assess their understanding of the material. Assess their accuracy and fluency, mark the words they say incorrectly, and listen for fluent reading. Use your judgment to determine whether students seem frustrated as they read. As a general guideline, students reading below 90% accuracy are likely to feel frustrated as they read. There are also a variety of published reading assessment tools that can be used to assess students' reading levels with the running record format.
- **Refer to other resources:** Other ways to determine instructional reading levels include checking your students' Individualized Education Plans, asking the school's resource teachers, or reviewing test scores. All of these resources should be able to give you the further information you need to determine at which reading level to begin your students.

Teachers can also use the texts in this series to scaffold the content for their students. At the beginning of the year, students at the lowest reading levels may need focused teacher guidance. As the year progresses, teachers can begin giving students multiple levels of the same text to allow them to work independently to improve their comprehension. This means each student would have a copy of the text at his or her independent reading level and instructional reading level. As students read the instructional-level texts, they can use the lower texts to better understand the difficult vocabulary. By scaffolding the content in this way, teachers can support students as they move up through the reading levels. This will encourage students to work with texts that are closer to the grade level at which they will be tested.

General Information About the Student Populations

Below-Grade-Level Students

By Dennis Benjamin

Gone are the days of a separate special education curriculum. Federal government regulations require that special needs students have access to the general education curriculum. For the vast majority of special needs students today, their Individualized Education Plans (IEPs) contain current and targeted performance levels but few short-term content objectives. In other words, the special needs students are required to learn the same content as their on-grade-level peers.

Be well aware of the accommodations and modifications written in students' IEPs. Use them in your teaching and assessment so they become routine. If you hold high expectations of success for all of your students, their efforts and performances will rise as well. Remember the root word of disability is ability. Go to the root needs of the learner and apply good teaching. The results will astound and please both of you.

English Language Learners

By Marcela von Vacano

Many school districts have chosen the inclusion model to integrate English language learners into mainstream classrooms. This model has its benefits as well as its drawbacks. One benefit is that English language learners may be able to learn from their peers by hearing and using English more frequently. One drawback is that these second-language learners cannot understand academic language and concepts without special instruction. They need sheltered instruction to take the first steps toward mastering English. In an inclusion classroom, the teacher may not have the time or necessary training to provide specialized instruction for these learners.

Acquiring a second language is a lengthy process that integrates listening, speaking, reading, and writing. Students who are newcomers to the English language are not able to process information until they have mastered a certain number of structures and vocabulary words. Students may learn social language in one or two years. However, academic language takes up to eight years for most students.

Teaching academic language requires good planning and effective implementation. Pacing, or the rate at which information is presented, is another important component in this process. English language learners need to hear the same word in context several times, and they need to practice structures to internalize the words. Reviewing and summarizing what was taught are absolutely necessary for English language learners.

General Information About the Student Populations *(cont.)*

On-Grade-Level Students

By Wendy Conklin

Often, on-grade-level students get overlooked when planning curriculum. More emphasis is usually placed on those who struggle and, at times, on those who excel. Teachers spend time teaching basic skills and even go below grade level to ensure that all students are up to speed. While this is a noble thing and is necessary at times, in the midst of it all, the on-grade-level students can get lost in the shuffle. We must not forget that differentiated strategies are good for the on-grade-level students, too. Providing activities that are too challenging can frustrate these students; on the other hand, assignments that are too easy can be boring and a waste of their time. The key to reaching this population successfully is to find just the right level of activities and questions while keeping a keen eye on their diverse learning styles.

Above-Grade-Level Students

By Wendy Conklin

In recent years, many state and school district budgets have cut funding that has in the past provided resources for their gifted and talented programs. The push and focus of schools nationwide is proficiency. It is important that students have the basic skills to read fluently, solve math problems, and grasp science concepts. As a result, funding has been redistributed in hopes of improving test scores on state and national standardized tests. In many cases, the attention has focused only on improving low test scores to the detriment of the gifted students who need to be challenged.

Differentiating the products you require from your students is a very effective and fairly easy way to meet the needs of gifted students. Actually, this simple change to your assignments will benefit all levels of students in your classroom. While some students are strong verbally, others express themselves better through nonlinguistic representation. After reading the texts in this book, students can express their comprehension through different means, such as drawings, plays, songs, skits, or videos. It is important to identify and address different learning styles. By giving more open-ended assignments, you allow for more creativity and diversity in your classroom. These differentiated products can easily be aligned with content standards. To assess these standards, use differentiated rubrics.

Strategies for Using the Leveled Texts

Below-Grade-Level Students

By Dennis Benjamin

Vocabulary Scavenger Hunt

A valuable prereading strategy is a Vocabulary Scavenger Hunt. Students preview the text and highlight unknown words. Students then write the words on specially divided pages. The pages are divided into quarters with the following headings: *Definition*, *Sentence*, *Examples*, and *Nonexamples*. A section called *Picture* is put over the middle of the chart.

Example Vocabulary Scavenger Hunt

astronomer

Definition a scientist who studies the universe and the objects within it	**Sentence** Astronomers use telescopes to discover new planets.
Examples Nicholas Copernicus; Galileo Galilei; Carl Sagan	**Nonexamples** George Washington; Ludwig van Beethoven; Rosa Parks

This encounter with new vocabulary enables students to use it properly. The definition identifies the word's meaning in student-friendly language. The sentence should be written so that the word is used in context. This helps the student make connections with background knowledge. Illustrating the sentence gives a visual clue. Examples help students prepare for factual questions from the teacher or on standardized assessments. Nonexamples help students prepare for ***not*** and ***except for*** test questions such as "All of these are explorers *except for…*" and "Which of these people is *not* an explorer?" Any information the student was unable to record before reading can be added after reading the text.

Strategies for Using the Leveled Texts *(cont.)*

Below-Grade-Level Students *(cont.)*

Graphic Organizers to Find Similarities and Differences

Setting a purpose for reading content focuses the learner. One purpose for reading can be to identify similarities and differences. This is a skill that must be directly taught, modeled, and applied. The authors of *Classroom Instruction That Works* state that identifying similarities and differences "might be considered the core of all learning" (Marzano, Pickering, and Pollock 2001, 14). Higher-level tasks include comparing and classifying information and using metaphors and analogies. One way to scaffold these skills is through the use of graphic organizers, which help students focus on the essential information and organize their thoughts.

Example Classifying Graphic Organizer

Astronaut/ Cosmonaut	Nation	Major Space Achievement	Date of Achievement	Spacecraft/ Mission
Yuri Gagarin	Soviet Union	First person to travel in space	April 12, 1961	*Vostok 1*
Alan Shepard	United States	First American in space	May 5, 1961	*Freedom 7*
Alexei A. Leonov	Soviet Union	First spacewalk	March 18, 1965	*Voskhod 2*
Neil A. Armstrong	United States	First person on the moon	July 16, 1969	*Apollo 11*

The Riddles Graphic Organizer allows students to compare and contrast the astronauts using riddles. Students first complete a chart you've designed. Then, using that chart, they can write summary sentences. They do this by using the riddle clues and reading across the chart. Students can also read down the chart and write summary sentences. With the chart below, students could write the following sentences: Gagarin and Leonov represented the Soviet Union in space. Neil Armstrong and Alan Shepard walked on the moon.

Example Riddles Graphic Organizer

Who am I?	Gagarin	Shepard	Leonov	Armstrong
I walked on the moon.		x		x
I represented the Soviet Union in space.	x		x	
I was the first person from my nation to explore space.	x	x		
I was a pilot before I became an astronaut.	x	x	x	x
I went into space in 1961.	x	x		

Strategies for Using the Leveled Texts *(cont.)*

Below-Grade-Level Students *(cont.)*

Framed Outline

This is an underused technique that bears great results. Many below-grade-level students have problems with reading comprehension. They need a framework to help them attack the text and gain confidence in comprehending the material. Once students gain confidence and learn how to locate factual information, the teacher can fade out this technique.

There are two steps to successfully using this technique. First, the teacher writes cloze sentences. Second, the students complete the cloze activity and write summary sentences.

Example Framed Outline

> On July 21, 1969, the first _______________ walked on the moon. His name was Neil _______________ . He and astronaut Edwin E. "Buzz" Aldrin Jr. spent more than two hours _______________ on the moon. They wore bulky _______________ .

Summary Sentences:

> On July 21, 1969, U.S. astronauts Neil Armstong and Edwin E. Adrin Jr. became the first humans to step foot on the moon. The Apollo 11 astronauts trained for years before becoming space pioneers.

Modeling Written Responses

A frequent criticism heard by special educators is that below-grade-level students write poor responses to content-area questions. This problem can be remedied if resource and classroom teachers model what good answers look like. While this may seem like common sense, few teachers take the time to do this. They just assume all children know how to respond in writing.

This is a technique you may want to use before asking your students to respond to the comprehension questions associated with the leveled texts in this series. First, read the question aloud. Then, write the question on an overhead and talk aloud about how you would go about answering the question. Next, write the answer using a complete sentence that accurately answers the question. Repeat the procedure for several questions so that students make the connection that quality written responses are your expectation.

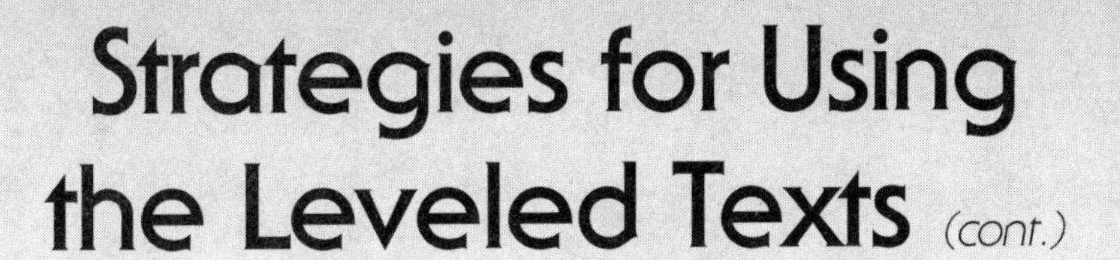

Strategies for Using the Leveled Texts *(cont.)*

English Language Learners

By Marcela von Vacano

Effective teaching for English language learners requires effective planning. In order to achieve success, teachers need to understand and use a conceptual framework to help them plan lessons and units. There are six major components to any framework. Each is described in more detail below.

1. Select and Define Concepts and Language Objectives—Before having students read one of the texts in this book, the teacher must first choose a science concept and language objective (reading, writing, listening, or speaking) appropriate for the grade level. Then, the next step is to clearly define the concept to be taught. This requires knowledge of the subject matter, alignment with local and state objectives, and careful formulation of a statement that defines the concept. This concept represents the overarching idea. The science concept should be written on a piece of paper and posted in a visible place in the classroom.

By the definition of the concept, post a set of key language objectives. Based on the content and language objectives, select essential vocabulary from the text. The number of new words selected should be based on students' English language levels. Post these words on a word wall that may be arranged alphabetically or by themes.

2. Build Background Knowledge—Some English language learners may have a lot of knowledge in their native language, while others may have little or no knowledge. The teacher will want to build the background knowledge of the students using different strategies such as the following:

Visuals: Use posters, photographs, postcards, newspapers, magazines, drawings, and video clips of the topic you are presenting. The texts in this series include multiple primary sources for your use.

Realia: Bring real-life objects to the classroom. If you are teaching about the plant life cycle, bring in items such as soil, seeds, roots, leaves, and flowers.

Vocabulary and Word Wall: Introduce key vocabulary in context. Create families of words. Have students draw pictures that illustrate the words and write sentences about the words. Also be sure you have posted the words on a word wall in your classroom.

Desk Dictionaries: Have students create their own desk dictionaries using index cards. On one side, they should draw a picture of the word. On the opposite side, they should write the word in their own language and in English.

Strategies for Using the Leveled Texts *(cont.)*

English Language Learners *(cont.)*

3. Teach Concepts and Language Objectives—The teacher must present content and language objectives clearly. He or she must engage students using a hook and must pace the delivery of instruction, taking into consideration students' English language levels. The concept or concepts to be taught must be stated clearly. Use the first languages of the students whenever possible or assign other students who speak the same languages to mentor and to work cooperatively with the English language learners.

Lev Semenovich Vygotsky, a Russian psychologist, wrote about the Zone of Proximal Development (ZPD). This theory states that good instruction must fill the gap that exists between the present knowledge of a child and the child's potential. Scaffolding instruction is an important component when planning and teaching lessons. English language learners cannot jump stages of language and content development. You must determine where the students are in the learning process and teach to the next level using several small steps to get to the desired outcome. With the leveled texts in this series and periodic assessment of students' language levels, teachers can support students as they climb the academic ladder.

4. Practice Concepts and Language Objectives—English language learners need to practice what they learn with engaging activities. Most people retain knowledge best after applying what they learn to their own lives. This is definitely true for English language learners. Students can apply content and language knowledge by creating projects, stories, skits, poems, or artifacts that show what they learned. Some activities should be geared to the right side of the brain, like those listed above. For students who are left-brain dominant, activities such as defining words and concepts, using graphic organizers, and explaining procedures should be developed. The following teaching strategies are effective in helping students practice both language and content:

> **Simulations**: Students learn by doing. For example, when teaching about the plant life cycle, you can have students figure out what they would need to grow a plant. First, they need to make a list and collect the necessary items, such as a clay pot or empty milk jug, planting soil, seeds, and water. They can fill the pot or jug with soil and plant a seed. They will need to water the plant daily, and make sure it gets enough sun. Lastly, students can measure and record how much the plant grows in a week, two weeks, or one month.
>
> **Literature response**: Read a text from this book. Have students choose two people described or introduced in the text. Ask students to create a conversation the people might have. Or, you can have students write journal entries about events in the daily lives of the famous scientists.

Strategies for Using the Leveled Texts *(cont.)*

English Language Learners *(cont.)*

4. Practice Concepts and Language Objectives *(cont.)*

Have a short debate: Make a controversial statement such as, "It isn't necessary for humans to explore space." After reading a text in this book, have students think about the question and take a position. As students present their ideas, one student can act as a moderator.

Interview: Students may interview a member of the family or a neighbor in order to obtain information regarding a topic from the texts in this book. For example: What was the reaction when Apollo 11 astronauts walked on the moon?

5. Evaluation and Alternative Assessments—We know that evaluation is used to inform instruction. Students must have the opportunity to show their understanding of concepts in different ways and not only through standard assessments. Use both formative and summative assessment to ensure that you are effectively meeting your content and language objectives. Formative assessment is used to plan effective lessons for a particular group of students. Summative assessment is used to find out how much the students have learned. Other authentic assessments that show day-to-day progress are: text retelling, teacher rating scales, student self-evaluations, cloze testing, holistic scoring of writing samples, performance assessments, and portfolios. Periodically assessing student learning will help you ensure that students continue to receive the correct levels of texts.

6. Home-School Connection—The home-school connection is an important component in the learning process for English language learners. Parents are the first teachers, and they establish expectations for their children. These expectations help shape the behavior of their children. By asking parents to be active participants in the education of their children, students get a double dose of support and encouragement. As a result, families become partners in the education of their children and chances for success in your classroom increase.

You can send home copies of the texts in this series for parents to read with their children. You can even send multiple levels to meet the needs of your second-language parents as well as your students. In this way, you are sharing your science content standards with your whole second-language community.

Strategies for Using the Leveled Texts *(cont.)*

Above-Grade-Level Students

By Wendy Conklin

Open-Ended Questions and Activities

Teachers need to be aware of activities that provide a ceiling that is too low for gifted students. When given activities like this, gifted students become bored. We know these students can do more, but how much more? Offering open-ended questions and activities will give high-ability students the opportunities to perform at or above their ability levels. For example, ask students to evaluate scientific topics described in the texts, such as: "Do you think the United States should be continuing space exploration?" or "What do you think our government should do to deal with global warming?" These questions require students to form opinions, think deeply about the issues, and form pro and con statements in their minds. To questions like these, there really is not one right answer.

The generic, open-ended question stems listed below can be adapted to any topic. There is one leveled comprehension question for each text in this book. These question stems can be used to develop further comprehension questions for the leveled texts.

- In what ways did…
- How might you have done this differently…
- What if…
- What are some possible explanations for…
- How does this affect…
- Explain several reasons why…
- What problems does this create…
- Describe the ways…
- What is the best…
- What is the worst…
- What is the likelihood…
- Predict the outcome…
- Form a hypothesis…
- What are three ways to classify…
- Support your reason…
- Compare this to modern times…
- Make a plan for…
- Propose a solution…
- What is an alternative to…

Strategies for Using the Leveled Texts *(cont.)*

Above-Grade-Level Students *(cont.)*

Student-Directed Learning

Because they are academically advanced, above-grade-level students are often the leaders in classrooms. They are more self-sufficient learners, too. As a result, there are some student-directed strategies that teachers can employ successfully with these students. Remember to use the texts in this book as jumpstarts so that students will be interested in finding out more about the science concepts presented. Above-grade-level students may enjoy any of the following activities:

- Writing their own questions, exchanging their questions with others, and grading the responses.
- Reviewing the lesson and teaching the topic to another group of students.
- Reading other nonfiction texts about these science concepts to further expand their knowledge.
- Writing the quizzes and tests to go along with the texts.
- Creating illustrated time lines to be displayed as visuals for the entire class.
- Putting together multimedia presentations about the scientific breakthroughs and concepts.

Tiered Assignments

Teachers can differentiate lessons by using tiered assignments, or scaffolded lessons. Tiered assignments are parallel tasks designed to have varied levels of depth, complexity, and abstractness. All students work toward one goal, concept, or outcome, but the lesson is tiered to allow for different levels of readiness and performance levels. As students work, they build on their prior knowledge and understanding. Students are motivated to be successful according to their own readiness and learning preferences.

Guidelines for writing tiered lessons include the following:

1. Pick the skill, concept, or generalization that needs to be learned.
2. Think of an on-grade-level activity that teaches this skill, concept, or generalization.
3. Assess the students using classroom discussions, quizzes, tests, or journal entries and place them in groups.
4. Take another look at the activity from Step 2. Modify this activity to meet the needs of the below-grade-level and above-grade-level learners in the class. Add complexity and depth for the above-grade-level students. Add vocabulary support and concrete examples for the below-grade-level students.

How to Use This Product

Readability Chart

Title of the Text	Star	Circle	Square	Triangle
Atoms	2.2	3.0	5.1	6.5
Elements, Molecules, and Mixtures	2.2	3.4	5.0	7.1
States of Matter	2.1	3.3	5.2	6.5
The Periodic Table	2.2	3.2	5.2	6.5
Chemical Reactions	2.0	3.5	5.1	6.5
Energy	2.2	3.5	4.5	6.5
Heat	1.8	3.5	4.5	6.5
Sunlight	1.5	3.4	4.8	6.6
Electrical Circuits	1.7	3.5	4.6	6.9
Vibrations	2.1	3.5	4.5	6.5
Radiant Light	2.2	3.2	5.0	6.5
Gravity	2.2	3.1	4.5	6.6
Relativity	2.2	3.0	4.9	6.5
Electromagnetism	2.2	3.5	5.2	6.7
Newton's Laws of Motion	2.0	3.4	4.6	6.5

Components of the Product

Primary Sources

- Each level of text includes multiple primary sources. These documents, photographs, and illustrations add interest to the texts. The scientific images also serve as visual support for second language learners. They make the texts more context rich and bring the texts to life.

How to Use This Product *(cont.)*

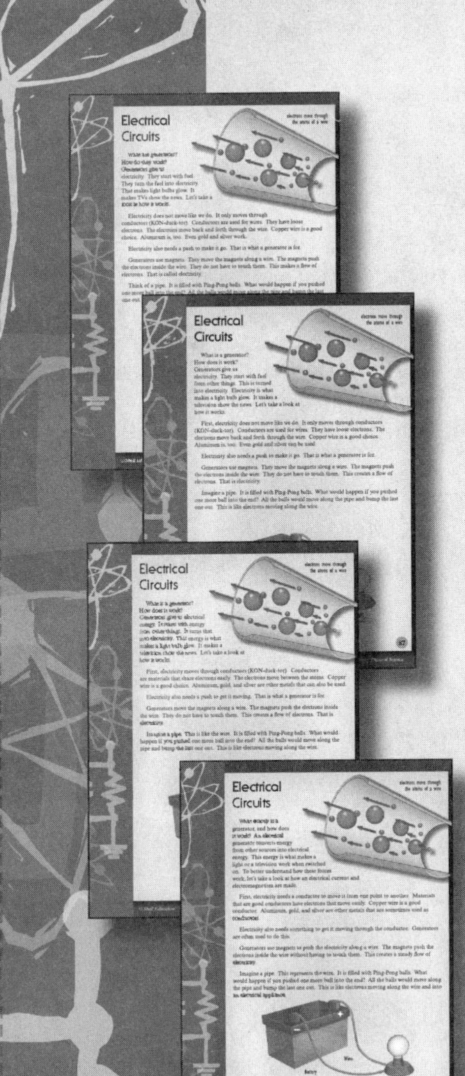

Components of the Product *(cont.)*

Comprehension Questions

- Each level of text includes one comprehension question. Like the texts, the comprehension questions were leveled by an expert. They are written to allow all students to be successful within a whole-class discussion. The questions for the same topic are closely linked so that the teacher can ask a question on that topic and all students will be able to answer. The lowest-level students might focus on the facts, while the upper-level students can delve deeper into the meanings.
- Teachers may want to base their whole-class question on the square level questions. Those were the starting points for all the other leveled questions.

The Levels

- There are 15 topics in this book. Each topic is leveled to four different reading levels. The images and fonts used for each level within a topic look the same.
- Behind each page number, you'll see a shape. These shapes indicate the reading levels of each piece so that you can make sure students are working with the correct texts. The reading levels fall into the ranges indicated to the left. See the chart at left for specific levels of each.

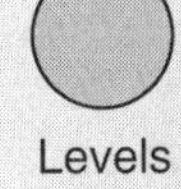

Levels 1.5–2.2

Levels 3.0–3.5

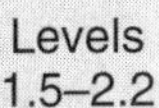

Levels 4.5–5.2

Levels 6.5–7.2

Leveling Process

- The texts in this series are taken from the *Science Readers* kits published by Teacher Created Materials. A reading expert went through the texts and leveled each one to create four distinct reading levels.
- After that, a special education expert and an English language learner expert carefully reviewed the lowest two levels and suggested changes that would help their students comprehend the texts better.
- The texts were then leveled one final time to ensure the editorial changes made during the process kept them within the ranges described to the left.

How to Use This Product *(cont.)*

Tips for Managing the Product

How to Prepare the Texts

- When you copy these texts, be sure you set your copier to copy photographs. Run a few test pages and adjust the contrast as necessary. If you want the students to be able to appreciate the images, you need to carefully prepare the texts for them.
- You also have full-color versions of the texts provided in PDF form on the CD. (See page 144 for more information.) Depending on how many copies you need to make, printing the full-color versions and copying those might work best for you.
- Keep in mind that you should copy two-sided to two-sided if you pull the pages out of the book. The shapes behind the page numbers will help you keep the pages organized as you prepare them.

Distributing the Texts

Some teachers wonder about how to hand the texts out within one classroom. They worry that students will feel insulted if they do not get the same papers as their neighbors. The first step in dealing with these texts is to set up your classroom as a place where all students learn at their individual instructional levels. Making this clear as a fact of life in your classroom is key. Otherwise, the students may constantly ask about why their work is different. You do not need to get into the technicalities of the reading levels. Just state it as a fact that every student will not be working on the same assignment every day. If you do this, then passing out the varied levels is not a problem. Just pass them to the correct students as you circle the room.

If you would rather not have students openly aware of the differences in the texts, you can try these ways to pass out the materials:

- Make a pile in your hands from star to triangle. Put your finger between the circle and square levels. As you approach each student, you pull from the top (star), above your finger (circle), below your finger (square), or the bottom (triangle). If you do not hesitate too much in front of each desk, the students will probably not notice.
- Begin the class period with an opening activity. Put the texts in different places around the room. As students work quietly, circulate and direct students to the right locations for retrieving the texts you want them to use.
- Organize the texts in small piles by seating arrangement so that when you arrive at a group of desks you have just the levels you need.

How to Use This Product *(cont.)*

Correlation to Standards

The No Child Left Behind (NCLB) legislation mandates that all states adopt academic standards that identify the skills students will learn in kindergarten through twelfth grade. While many states had already adopted academic standards prior to NCLB, the legislation set requirements to ensure the standards were detailed and comprehensive.

Standards are designed to focus instruction and guide adoption of curricula. Standards are statements that describe the criteria necessary for students to meet specific academic goals. They define the knowledge, skills, and content students should acquire at each level. Standards are also used to develop standardized tests to evaluate students' academic progress.

In many states today, teachers are required to demonstrate how their lessons meet state standards. State standards are used in the development of Shell Education products, so educators can be assured that they meet the academic requirements of each state.

How to Find Your State Correlations

Shell Education is committed to producing educational materials that are research and standards based. In this effort, all products are correlated to the academic standards of the 50 states, the District of Columbia, and the Department of Defense Dependent Schools. A correlation report customized for your state can be printed directly from the following website: **http://www.shelleducation.com**. If you require assistance in printing correlation reports, please contact Customer Service at 1-800-877-3450.

McREL Compendium

Shell Education uses the Mid-continent Research for Education and Learning (McREL) Compendium to create standards correlations. Each year, McREL analyzes state standards and revises the compendium. By following this procedure, they are able to produce a general compilation of national standards.

Each reading comprehension strategy assessed in this book is based on one or more McREL content standards. The following chart shows the McREL standards that correlate to each lesson used in the book. To see a state-specific correlation, visit the Shell Education website at **http://www.shelleducation.com**.

McREL	Benchmark	Text
8-III-1	Knows that matter is made up of tiny particles called atoms, and different arrangements of atoms into groups compose all substances	Atoms
8-III-2	Knows that atoms often combine to form a molecule (or crystal), the smallest particle of a substance that retains its properties	Elements, Compounds, and Mixtures
8-III-3	Knows that states of matter depend on molecular arrangement and motion (e.g., molecules in solids are packed tightly together and their movement is restricted to vibrations; molecules in liquids are loosely packed and move easily past each other; molecules in gases are quite far apart and move about freely)	States of Matter
8-III-4	Knows that substances containing only one kind of atom are elements and do not break down by normal laboratory reactions (e.g., heating, exposure to electric current, reaction with acids); over 100 different elements exist	Elements, Compounds, and Mixtures
8-III-5	Knows that many elements can be grouped according to similar properties (e.g., highly reactive metals, less-reactive metals, highly reactive nonmetals, almost completely nonreactive gases)	Periodic Table
8-III-6	Understands the conservation of mass in physical and chemical change (e.g., no matter how substances within a closed system interact with one another, the total weight of the system remains the same; the same number of atoms of a single element weighs the same, no matter how the atoms are arranged)	Chemical Reactions
8-III-7	Knows methods used to separate mixtures into their component parts (boiling, filtering, chromatography, screening)	Elements, Compounds, and Mixtures
8-III-8	Knows that substances react chemically in characteristic ways with other substances to form new substances (compounds) with different characteristic properties	Chemical Reactions
8-III-9	Knows factors that influence reaction rates (e.g., types of substances involved, temperature, concentration of reactant molecules, amount of contact between reactant molecules)	Chemical Reactions
8-III-10	Knows that oxidation is the loss of electrons, and commonly involves the combining of oxygen with another substance (e.g., the processes of burning and rusting)	Chemical Reactions
9-III-1	Knows that energy is a property of many substances (e.g., heat energy is in the disorderly motion of molecules and in radiation; chemical energy is in the arrangement of atoms; mechanical energy is in moving bodies or in elastically distorted shapes; electrical energy is in the attraction or repulsion between charges)	Energy
9-III-3	Knows that heat energy flows from warmer materials or regions to cooler ones through conduction, convection, and radiation	Heat
9-III-4	Knows how the Sun acts as a major source of energy for changes on the Earth's surface (i.e., the Sun loses energy by emitting light; some of this light is transferred to the Earth in a range of wavelengths including visible light, infrared radiation, and ultraviolet radiation)	Sunlight
9-III-5	Knows that electrical circuits provide a means of transferring electrical energy to produce heat, light, sound, and chemical changes	Electrical Circuits
9-III-7	Knows that vibrations (e.g., sounds, earthquakes) move at different speeds in different materials, have different wavelengths, and set up wave-like disturbances that spread away from the source	Vibrations
9-III-8	Knows ways in which light interacts with matter (e.g., transmission, including refraction; absorption; scattering, including reflection)	Light
10-III-1	Understands general concepts related to gravitational force (e.g., every object exerts gravitational force on every other object; this force depends on the mass of the objects and their distance from one another; gravitational force is hard to detect unless at least one of the objects, such as the Earth, has a lot of mass)	Gravity and Relativity
10-III-2	Knows that just as electric currents can produce magnetic forces, magnets can cause electric currents	Electro-magntism
10-III-3	Knows that an object's motion can be described and represented graphically according to its position, direction of motion, and speed	Newton's Laws of Motion
10-III-4	Understands effects of balanced and unbalanced forces on an object's motion (e.g., if more than one force acts on an object along a straight line, then the forces will reinforce or cancel one another, depending on their direction and magnitude; unbalanced forces such as friction will cause changes in the speed or direction on an object's motion)	Newton's Laws of Motion
10-III-5	Knows that an object that is not being subjected to a force will continue to move at a constant speed and in a straight line	Newton's Laws of Motion

Atoms

Think of your breakfast. Look at this page. What's the same between them? What's the same between this page and your hands? There is something that is the same about everything around you. Can you guess what it is?

All things are made of matter. Matter makes up everything. Matter is anything that takes up space. Even if the space is too small to see, matter is still there.

But what is matter made of? Matter is made of tiny parts called atoms. Atoms are the building blocks of everything. They are everywhere. They are the air you breathe. They are the food you eat. They are the things you touch. They are the clothes you wear. Atoms are every part of you.

Different Atoms, Different Things

There are many different things in the world. Each thing is made of different atoms. Do you know what an iron bar is made of? It is made of iron atoms. How about oxygen gas? That is easy. It is made of oxygen atoms.

Things have properties (PROP-er-tees). They get them from the atoms they are made of. Iron atoms let electricity flow through them. That means an iron bar does, too. Oxygen atoms grab other atoms. They do it all the time. Oxygen gas does the same thing.

Some things are different. They are made of more than one kind of atom. Rust is made of two kinds. It has both iron atoms and oxygen atoms. It is a mix of atoms. It has new properties.

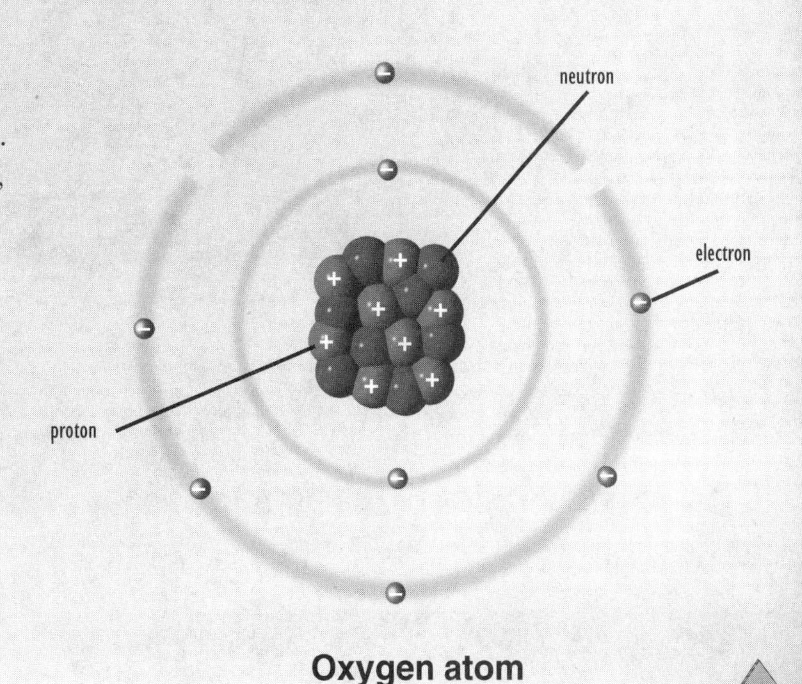

Oxygen atom

Inside the Atom

Look again. There are clouds. They are made of tiny things. They are called electrons (ee-LEK-trons). They zoom around the nucleus. The atom has protons. It has electrons. It always has the same number of electrons and protons.

Protons have charges. Electrons have charges. The proton charge is a plus. The electron charge is a minus. They all balance out. The whole atom is not a plus. It is not a minus. It has a balanced charge.

For a long time, scientists thought they knew all the parts of an atom. Then they learned that there is more inside. Protons and neutrons are made of even smaller pieces. They are called quarks (kwarks). Quarks are the smallest parts of an atom.

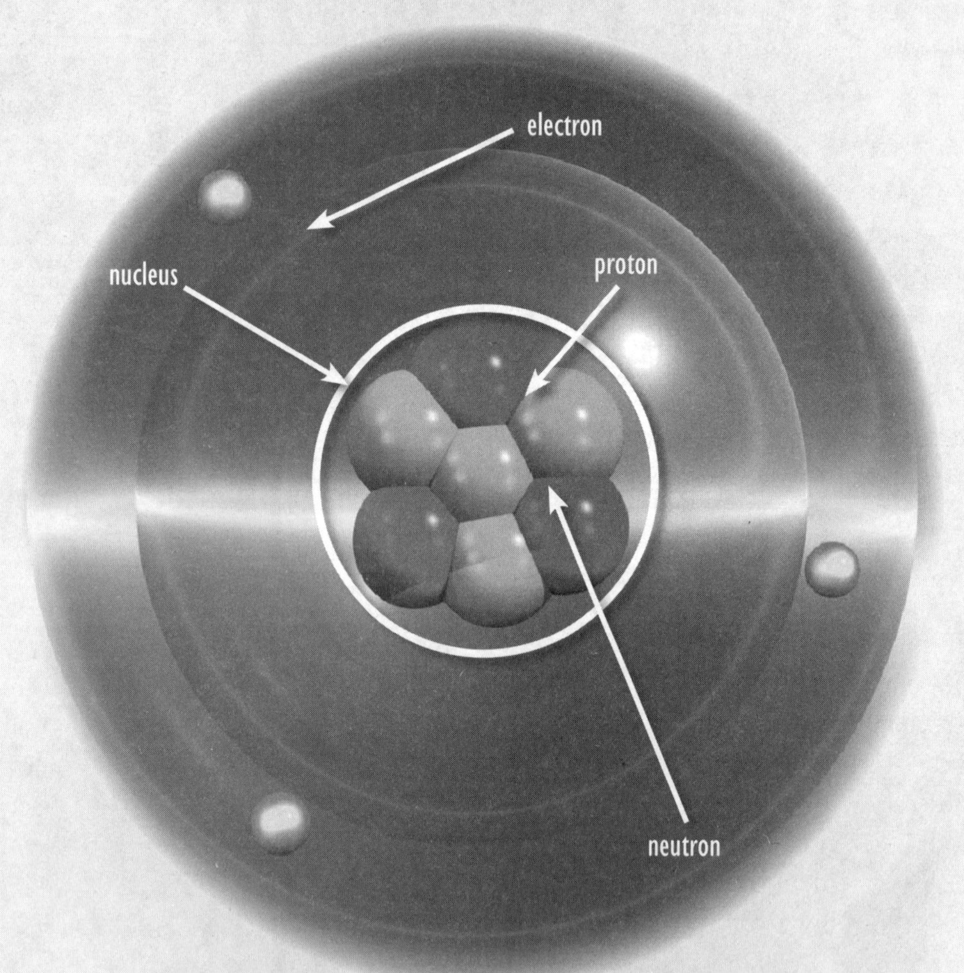

Comprehension Question

What are atoms?

Atoms

What's the same between this page and your breakfast? What's the same between this page and your hands? There is something that is the same about everything around you. Can you guess what it is?

All things are made of matter. Matter makes up everything. Matter is anything that takes up space. Even if the space is too small to see, matter is still there.

But what is matter made of? Matter is made of tiny particles. They are called atoms. Atoms are the building blocks of everything. They are everywhere. They are the air you breathe. They are the food you eat. They are the things you touch. They are the clothes you wear. Atoms are every part of you.

Different Atoms, Different Things

Different things are made of different atoms. Do you know what an iron bar is made of? It is made of iron atoms. How about oxygen gas? It is made of oxygen atoms.

Objects have properties (PROP-er-tees). They get them from the atoms they are made of. Iron atoms let electricity flow through them. That means an iron bar does, too. Oxygen atoms are quick to grab other atoms. Oxygen gas does the same thing.

Some objects are different. They are made of more than one kind of atom. Rust is made of two kinds. It has iron atoms and oxygen atoms. Because it is a mix of atoms, it has new properties.

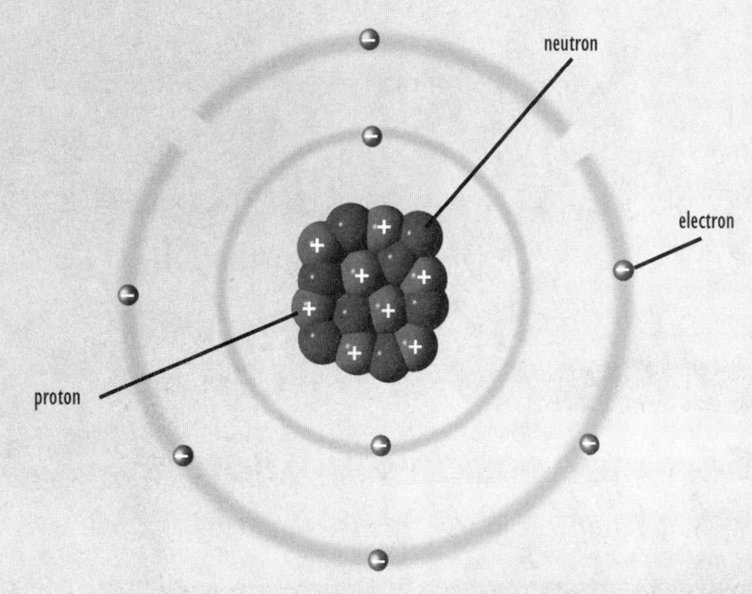

Oxygen atom

Inside the Atom

Let's take a peek inside an atom. There is a lot going on. First look at the small, tightly packed center. This is called the nucleus (NOO-klee-uhs). The nucleus is made of even tinier parts. Some are called protons (PROH-tons). Some are called neutrons (NOO-trons). There is about the same number of each. They are squeezed tightly together. They form the nucleus.

Look again. There are clouds of tiny particles. They are called electrons (ee-LEK-trons). They zoom around the nucleus. The atom always has the same number of electrons and protons.

Protons have charges. Electrons have charges. The proton charge is positive. The electron charge is negative. They all balance out. The whole atom is not positive. It is not negative. It has a balanced charge.

For a long time, scientists thought they knew all the parts of an atom. Then they learned that there is more inside. Protons and neutrons are made of even smaller pieces. They are called quarks (kwarks). Quarks are the smallest parts of an atom.

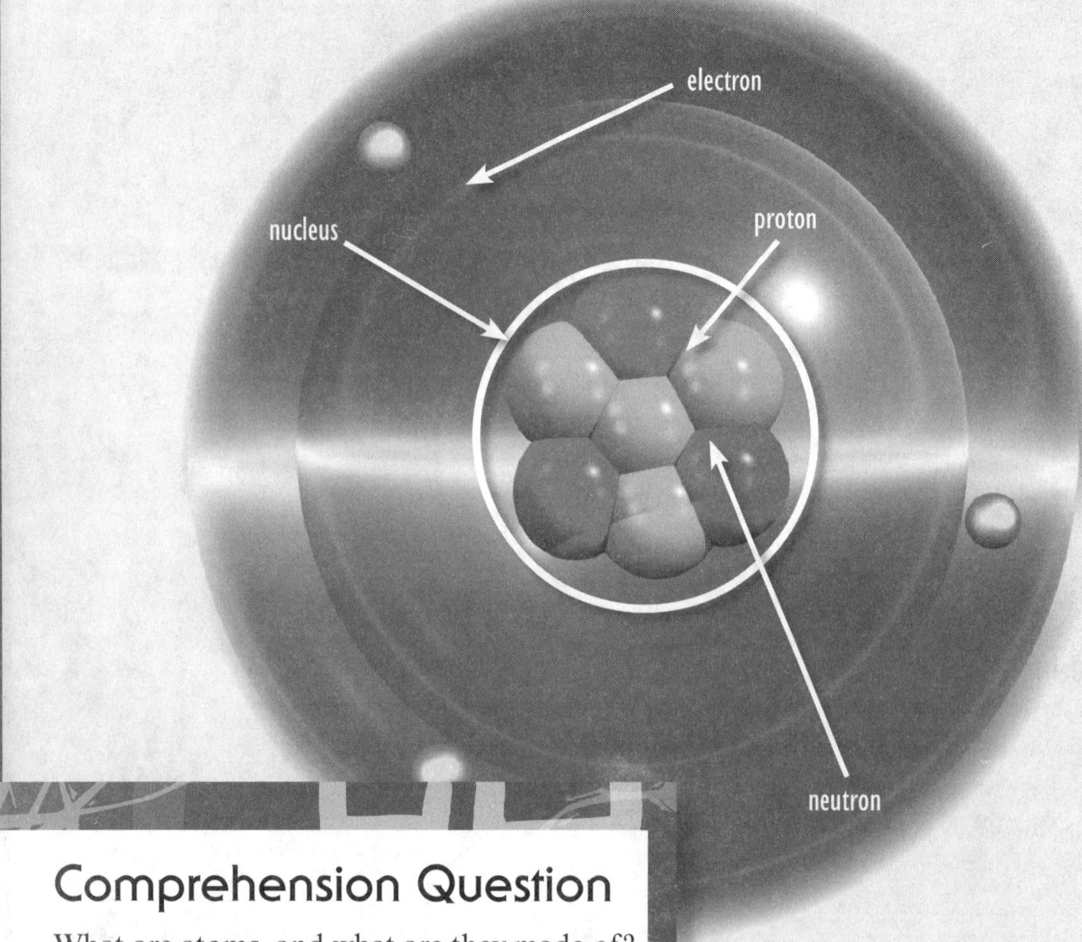

Comprehension Question

What are atoms, and what are they made of?

Atoms

This page has something important in common with the breakfast you ate this morning. Your breakfast has something in common with your hands holding this page. In fact, everything around you has it in common. Can you guess what it is?

All things are made of matter. Matter makes up everything. Scientists say that matter is anything that takes up space. Even if the space is too small to see, matter is still there.

But what is matter made of? No matter what, matter is made of tiny particles. They are called atoms. Atoms are the building blocks of everything. They are everywhere. They are the air you breathe. They are the food you eat, the things you touch, and the clothes you wear. They are every part of you.

Different Atoms, Different Things

Different things are made of different atoms. An iron bar is made of iron atoms. Oxygen gas is made of oxygen atoms.

Objects get properties (PROP-er-tees) from the atoms they are made of. Iron atoms conduct electricity. An iron bar, then, also conducts electricity. Oxygen atoms combine with other atoms easily. Oxygen gas does the same thing.

Some objects are made of more than one kind of atom. Rust is a combination of iron atoms and oxygen atoms. Because it is made of a mix of atoms, it has different properties.

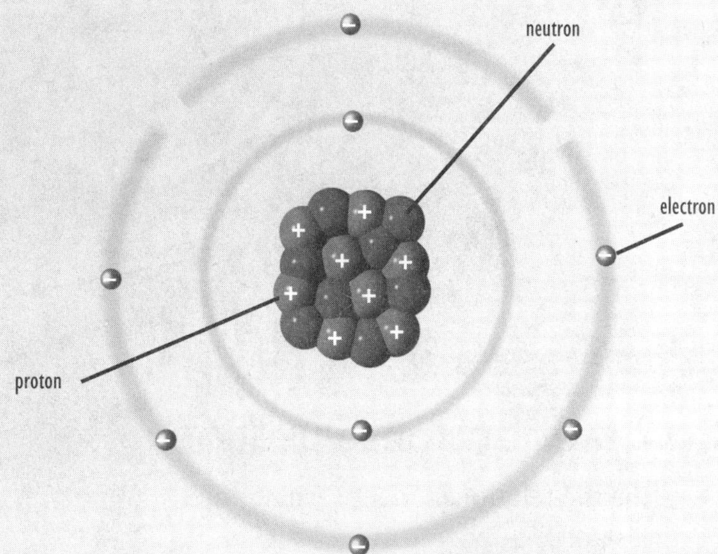

Oxygen atom

Inside the Atom

If you could peek inside an atom, you would see a lot going on. First you might notice its small, tightly-packed center. This is called the nucleus (NOO-klee-uhs). Inside the nucleus are even tinier particles. They are called protons (PROH-tons) and neutrons (NOO-trons). There are around the same number of protons and neutrons in an atom. Squeezed tightly together, they form the nucleus.

Look again, and you will see clouds of particles called electrons (uh-LEK-trons). They orbit around the nucleus. The same number of electrons and protons are in each atom.

Protons and electrons each have a charge. In protons, the charge is positive. In electrons, the charge is negative. Together, they balance the atom's charge. The atom as a whole is neither positive nor negative.

For a long time, scientists thought they knew all the parts of an atom. Then they learned that there is more. Protons and neutrons are made of something even smaller. They are tiny particles called quarks (kwarks). Quarks are the smallest parts of an atom.

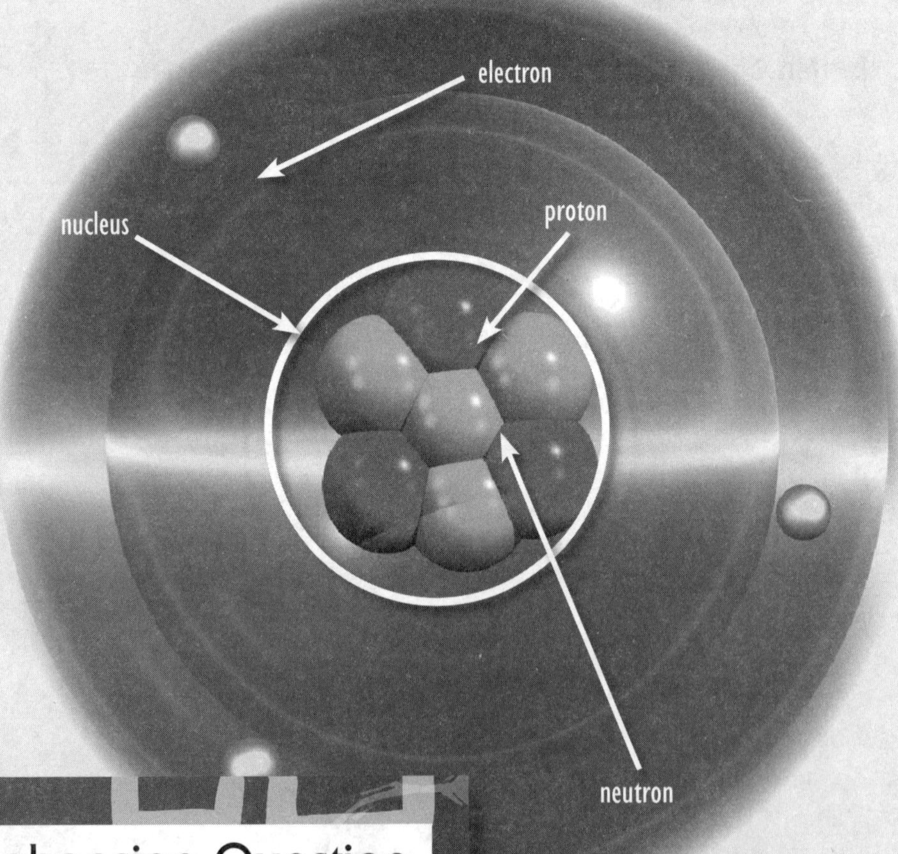

Comprehension Question

Describe the relationship between atomic particles and matter.

Atoms

This page has something important in common with the breakfast you ate this morning. Your breakfast has something in common with your hands holding this page. In fact, everything around you has it in common; can you guess what it is?

It's matter, and everything is composed of it. Matter is defined as anything that takes up space. Even microscopic substances take up space; therefore they are matter.

No matter what, matter is made of tiny particles called atoms. Atoms are the building blocks of everything in the universe. They are everything in the world around you: they are the air you breathe, the food you eat, the things you touch, and the clothes you wear. They even are the building blocks of you, your family, and your friends!

Different Atoms, Different Things

Different things are made of different atoms. An iron bar is made of iron atoms; oxygen gas is made of oxygen atoms.

Objects get properties (PROP-er-tees) from the atoms they are made of. For instance, iron atoms conduct electricity. An iron bar, then, also conducts electricity. Oxygen atoms combine with other atoms easily. Oxygen gas is one of the most reactive substances known to man.

Some substances are combinations of different atoms. Rust is composed of iron and oxygen atoms. Because it is made of a mix of atoms, it has different properties.

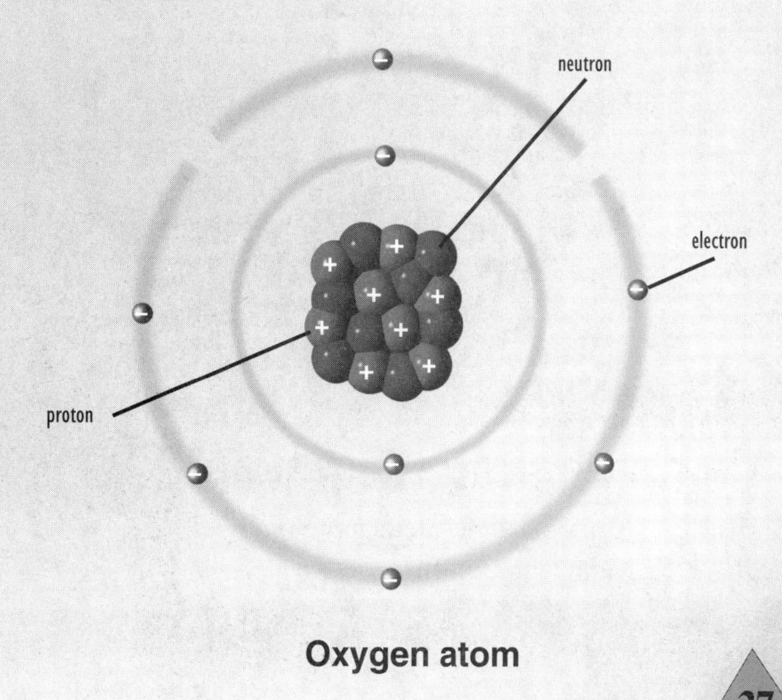

Oxygen atom

Inside the Atom

If you could peek inside an atom, you would see a lot going on. First you might notice its small, tightly-packed center: the nucleus (NOO-klee-uhs). Inside the nucleus are even tinier particles: protons (PROH-tons) and neutrons (NOO-trons). Usually, there is approximately the same number of protons and neutrons in the nucleus. They are squeezed tightly together; these forces are some of the strongest in the universe.

Look again; around the nucleus you will see clouds of particles called electrons (uh-LEK-trons). They orbit around the nucleus. Each atom has the same number of electrons and protons. If it doesn't, it tries to balance out.

Protons and electrons are charged particles. Protons have a positive charge; electrons have a negative charge. Since there are the same number of each, they balance the atom's charge. The atom as a whole is neither positive nor negative.

With the discovery of the electron, scientists thought they knew all the parts of an atom. In 1961, two scientists suggested there were even smaller particles. They were named quarks (kwarks), after a nonsense word from a novel, because the scientists weren't sure they really existed. Over the next twenty years, scientists found that protons and neutrons are made of these smallest particles.

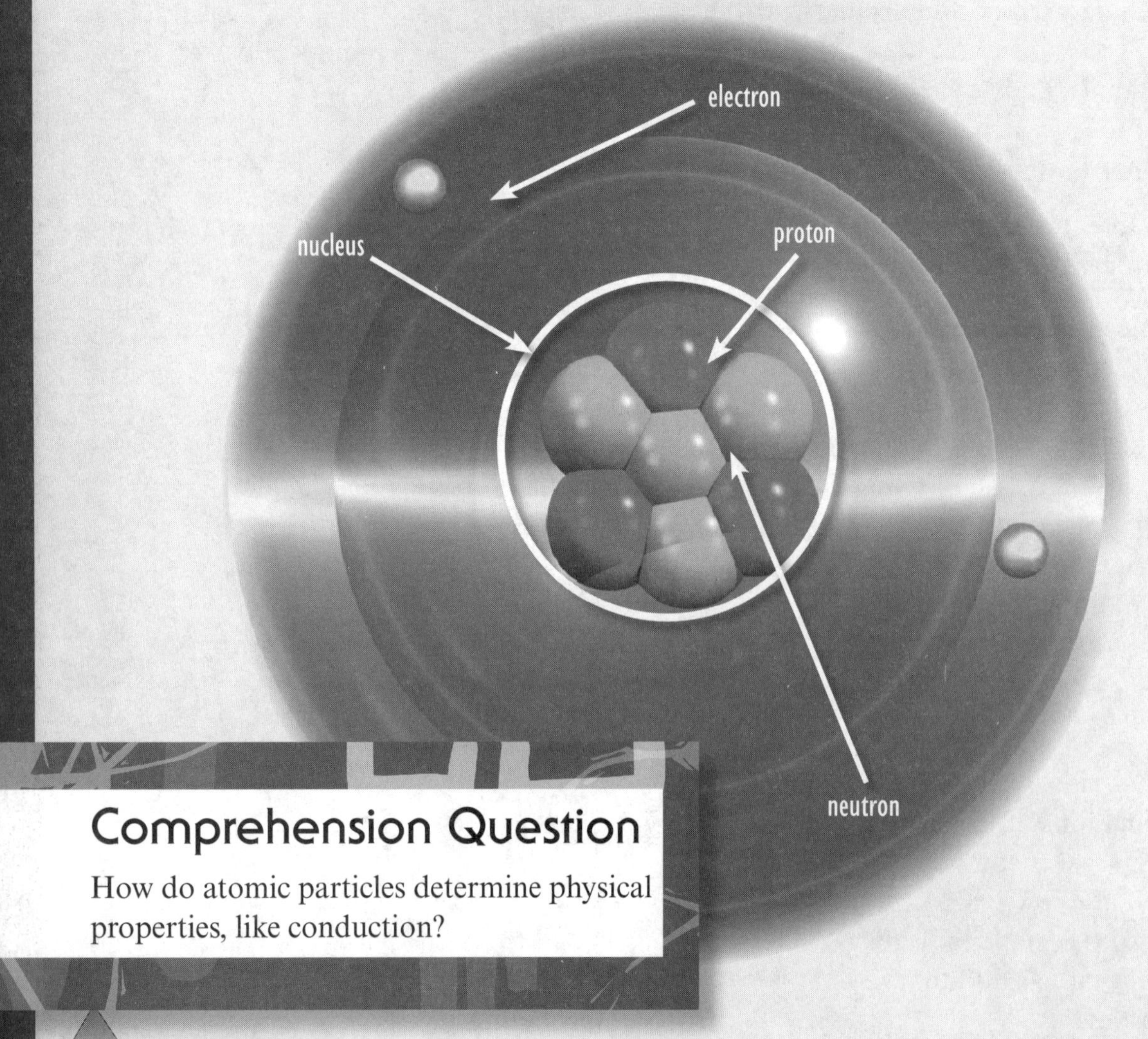

Comprehension Question

How do atomic particles determine physical properties, like conduction?

Elements, Molecules, and Mixtures

Atoms and Elements

Iron atom

proton

neutron

electron

All things are made of atoms. Atoms are tiny particles (PAR-tuh-kuhls). Your desk is made of atoms. Air is made of atoms. Even you are made of atoms. Atoms are very small. Do you know how many can fit in a teaspoon? Write a one. Write twenty-four zeroes after it. That many! No one can see them with just their eyes. It takes a strong microscope to see them.

A thing can be made of the same kind of atoms. That is called an element (ELL-em-ent). It is very hard to turn one element into another. Iron will always be iron. You can't turn it into other elements. You can heat it. You can hit it. You can drop it in acid. It doesn't matter what you do. It will still be iron. It may not look the same, but it will still be made of iron atoms.

There are about 100 types of atoms. They can be combined in many ways. Each way makes a different thing. This is called atomic arrangement (uh-TOM-ik uh-RANGE-muhnt).

Molecules and Compounds

Atoms can clump up with other atoms. That makes molecules (MOL-uh-kyools). A molecule has some atoms. They are stuck together. A lot of the same clumps is a compound (KOM-pownd). The compound is a new thing. It is different from the elements that make it.

Water is a compound. It is made from two kinds of atoms. It has two hydrogen atoms. It has one oxygen atom. Each water molecule has three atoms. This is written as H_2O. The H stands for hydrogen. The number two means that there are two of those atoms. The O stands for oxygen. No number after the O means there is just one atom. Water is made of oxygen and hydrogen. It isn't like either of them. It is a compound. Compounds are not like their parts.

How are compounds made? They are made of elements. Reactive (ree-AK-tiv) elements join with others to make compounds.

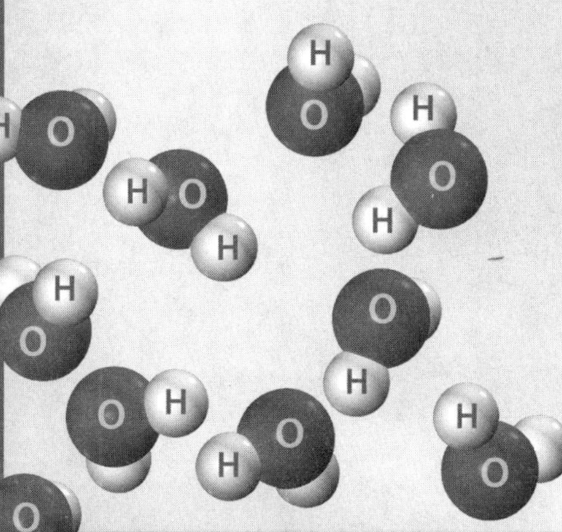

Water molecules

Mixtures

A mixture is not the same as a compound. Air and blood are mixtures. They have many types of atoms and molecules. Not all of the atoms are in compounds. They can be separated. You just need to know how.

You can pull apart a mixture in steps. You use the properties of the things in the mixture. These are things such as their melting and boiling points. It can be how magnets affect them. You can use magnets to pull out the magnetic parts. It can be the size of its solid chunks. You can use a filter that sorts the big chunks from the small chunks.

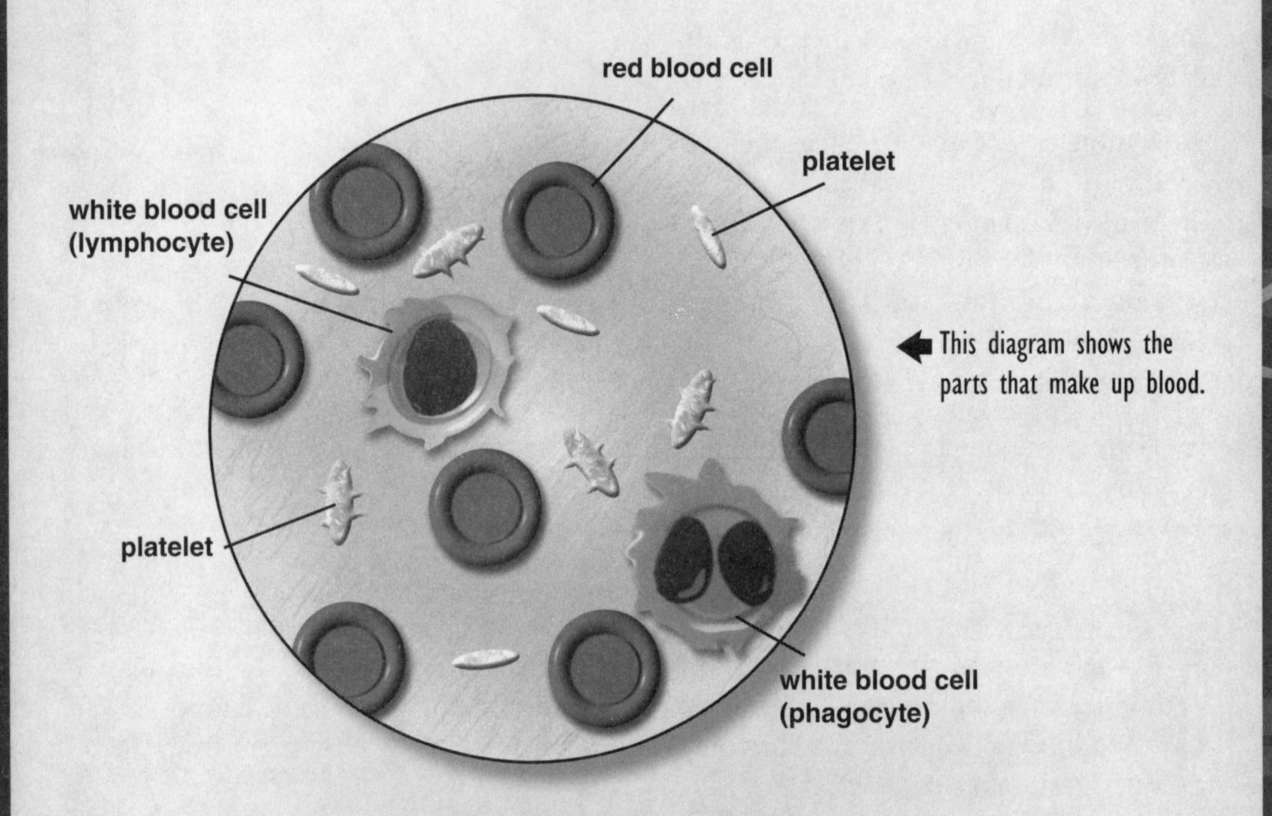

This diagram shows the parts that make up blood.

Comprehension Question

What are molecules made of?

Elements, Molecules, and Mixtures

Atoms and Elements

Iron atom

proton

neutron

electron

All matter is made of atoms. Atoms are tiny particles (PAR-tuh-kuhls). Even air is made of atoms. Atoms are small. A million billion billion of them fit in a teaspoon. No one can see them without help. It takes a strong microscope to see them.

A thing can be made of the same kind of atoms. That is called an element (ELL-em-ent). It is very hard to turn one element into another element. Iron will always be iron. You can't turn it into other elements. You can heat it. You can hit it. You can drop it in acid. It doesn't matter what you do. It will still be iron. It may not look the same. It will still be made of iron atoms.

There are about 100 types of atoms. They can be put together in many ways. Each way makes a different thing. This is called atomic arrangement (uh-TOM-ik uh-RANGE-muhnt).

Molecules and Compounds

Atoms join to make molecules (MOL-uh-kyools). A molecule has some atoms stuck together. That is a compound (KOM-pownd). The compound is different from the elements that make it.

Water is made from hydrogen and oxygen. But it isn't like either of them. Water is a compound. Each water molecule has two kinds of atoms. There are two hydrogen atoms and one oxygen atom. This is written as H_2O. The H stands for hydrogen. The number two means that there are two of those atoms. The O stands for oxygen. No number after the O means there is just one atom.

How are compounds made? They are made of reactive (ree-AK-tiv) elements. Reactive elements join with others. Some elements are very reactive. Some are not. Very reactive elements are very likely to join up to make compounds.

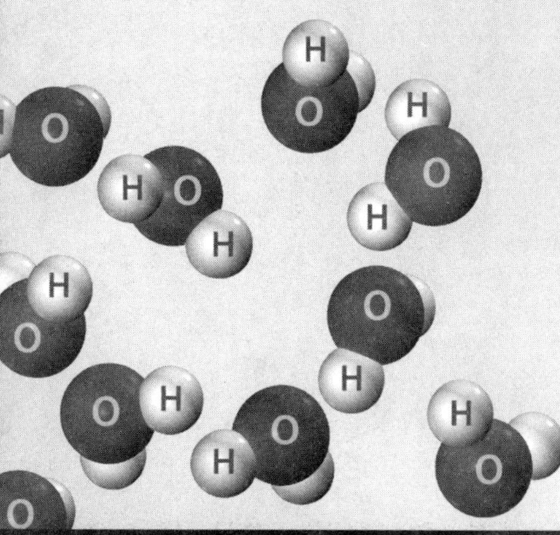

Water molecules

Mixtures

A mixture is not the same as a compound. Air and blood are mixtures. They have many types of atoms and molecules. Not all of the atoms are in compounds. They can be separated. You just need to know how.

You can separate a mixture in steps. You use the properties of the things in the mixture. These are things such as their melting and boiling points. It can be how magnets affect them. It can be the size of its solid chunks. You can use magnets to pull out the magnetic parts. You can use a filter that sorts the big chunks from the small chunks.

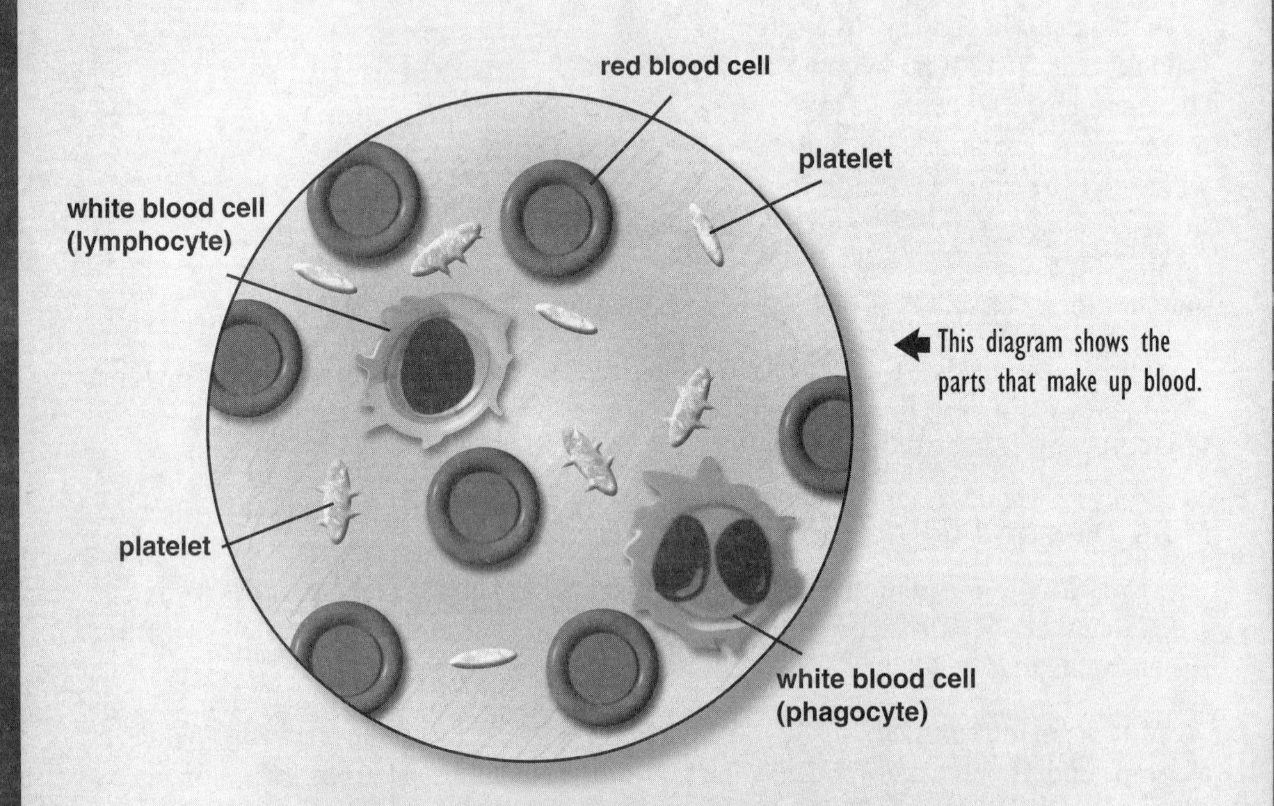

This diagram shows the parts that make up blood.

Comprehension Question

How are atoms and molecules related?

Elements, Molecules, and Mixtures

Atoms and Elements

Iron atom

proton

neutron

electron

All matter is made of atoms. Atoms are tiny particles (PAR-tuh-kuhls). Even air is made of atoms. Atoms are small. A million billion billion of them fit in a teaspoon. No one can see them without help. It takes a strong microscope to see them.

When something is made of the same kind of atoms, it is called an element (ELL-em-ent). It is very hard to turn one element into another element. In other words, iron will always be iron. You can't turn it into other elements. You can heat it. You can hit it. You can drop it in acid. It doesn't matter what you do. It will still be iron. It may not look the same, but it will still be made of iron atoms.

There are about 100 types of atoms. They can be put together in many ways. Each way makes something different. This is called atomic arrangement (uh-TOM-ik uh-RANGE-muhnt).

Molecules and Compounds

Atoms can join to make molecules (MOL-uh-kyools). A molecule has two or more atoms stuck together. They become a new substance called a compound (KOM-pownd). A compound has different properties from the elements that make it.

For example, water is made from hydrogen and oxygen. But it isn't like either of them. Water is a compound. Each water molecule has two kinds of atoms. There are two hydrogen and one oxygen atom. This is written as H_2O. The number two means that there are two hydrogen atoms in the molecule. No number after the O means there is just one atom of oxygen.

Compounds are made by reactive (ree-AK-tiv) elements. Reactive elements join easily with others. Some elements are very reactive. Some are not. The more reactive an element, the more likely it will form compounds.

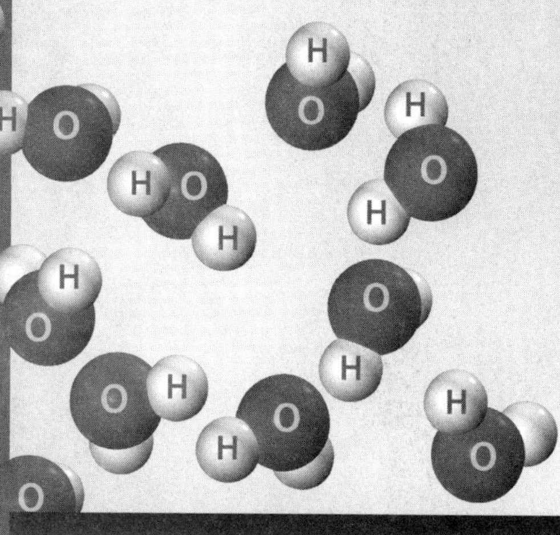

Water molecules

Mixtures

A mixture is not the same as a compound. Some everyday mixtures are air and blood. They contain many different types of atoms and molecules. Not all of the atoms and molecules are joined through reactions. They can be separated easily if you know how.

The way to separate a mixture is to use the properties of the substances in the mixture. These properties are things such as the melting and boiling points. Another is whether or not it is magnetic. And one more property is the size of its solid chunks. You can use magnets to separate out the magnetic molecules. You can use a sieve to separate the big chunks from the small chunks.

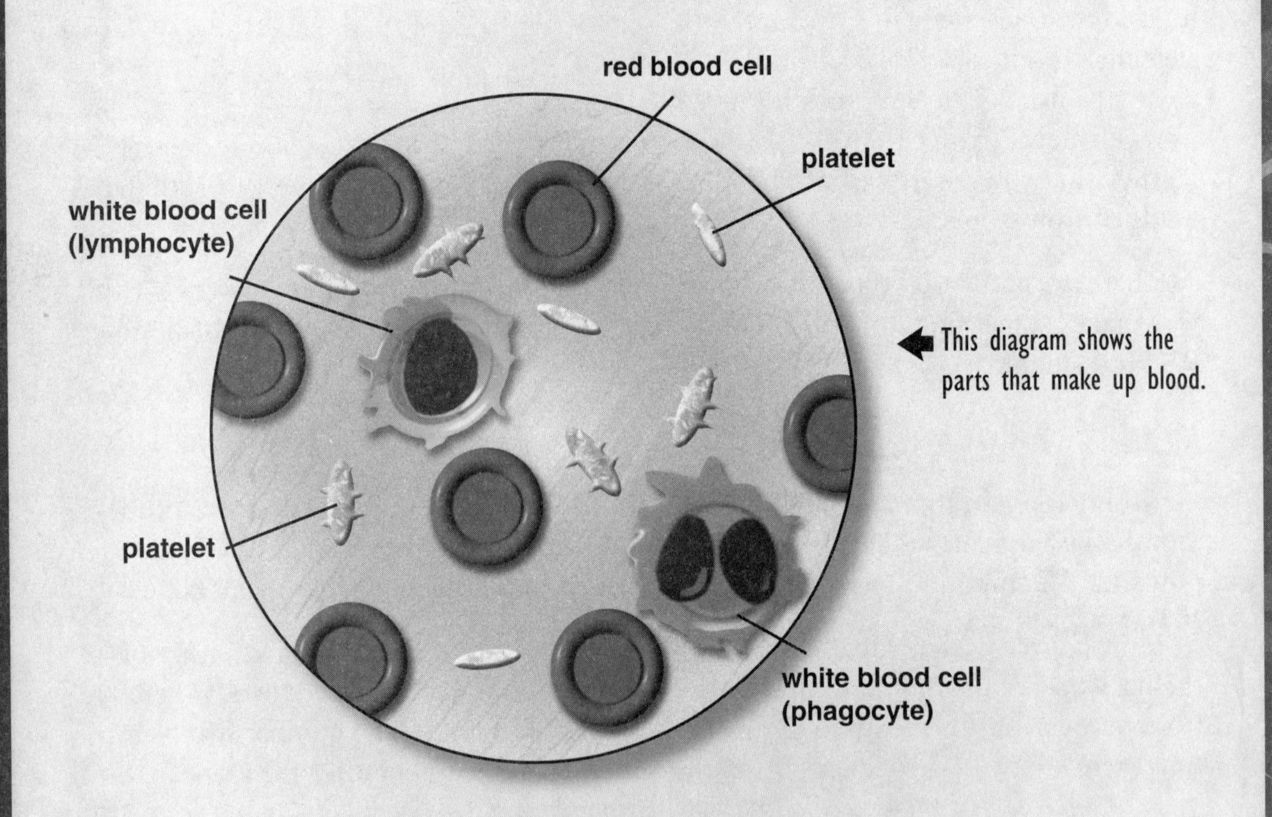

This diagram shows the parts that make up blood.

Comprehension Question

Compare and contrast atoms and molecules.

Elements, Molecules, and Mixtures

Atoms and Elements

Iron atom

proton

neutron

electron

Everything is made of atoms. Atoms are tiny particles (PAR-tuh-kuhls). Your desk, the air around you, even you are made of atoms. Atoms are so small that a million billion billion of them fit on a teaspoon. Humans can only see them with powerful electron microscopes.

When a substance is made entirely of the one kind of atom, it is called an element (ELL-em-ent). Elements cannot be easily transformed into other elements. In other words, iron will always be iron. You can heat it, hit it, or drop it in acid. No matter what you do, it will still be iron. It may not look the same after so much abuse, but it will still be composed of iron atoms.

There are about 100 different elements. They can be put together in many different ways. Each different way makes one of the millions of different things that exist. This is called atomic arrangement (uh-TOM-ik uh-RANGE-muhnt).

Molecules and Compounds

Atoms combine to make molecules (MOL-uh-kyools). In molecules, the component atoms share electrons. Together, they become a new substance called a compound (KOM-pownd). A compound has different properties from the elements that went into it.

For example, water is a compound of hydrogen and oxygen, but it isn't like either of them. Each water molecule has two hydrogen and one oxygen atom. The atomic arrangement is written as H_2O. The number two means that there are two hydrogen atoms in the molecule; no number after the O means there is just one oxygen.

Compounds are made by reactive (ree-AK-tiv) elements that join easily with others in chemical reactions. Some elements are very reactive; some are not. The more reactive an element, the more likely it will form compounds.

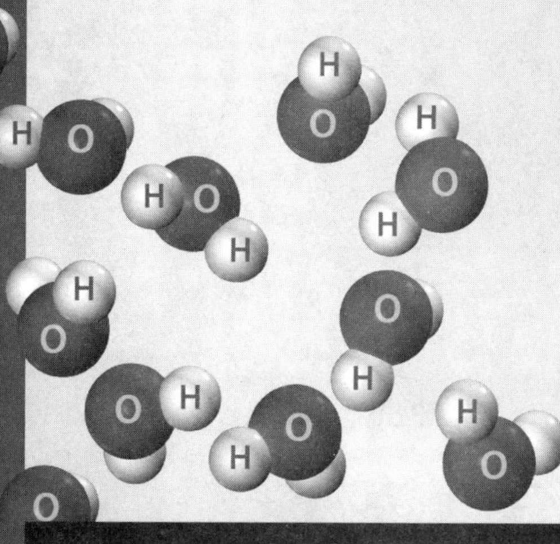

Water molecules

Mixtures

A mixture is not the same as a compound. Some everyday mixtures are air and blood. They contain many different types of atoms and molecules, and not all of them are the product of chemical reactions. They can be separated easily if you know how.

To separate a mixture, use the properties of the substances in the mixture such as their melting and boiling points. Magnetic ingredients or ingredients that form solid chunks are easy to separate out. Magnets can separate out the magnetic molecules. A sieve can separate the big chunks from the small chunks.

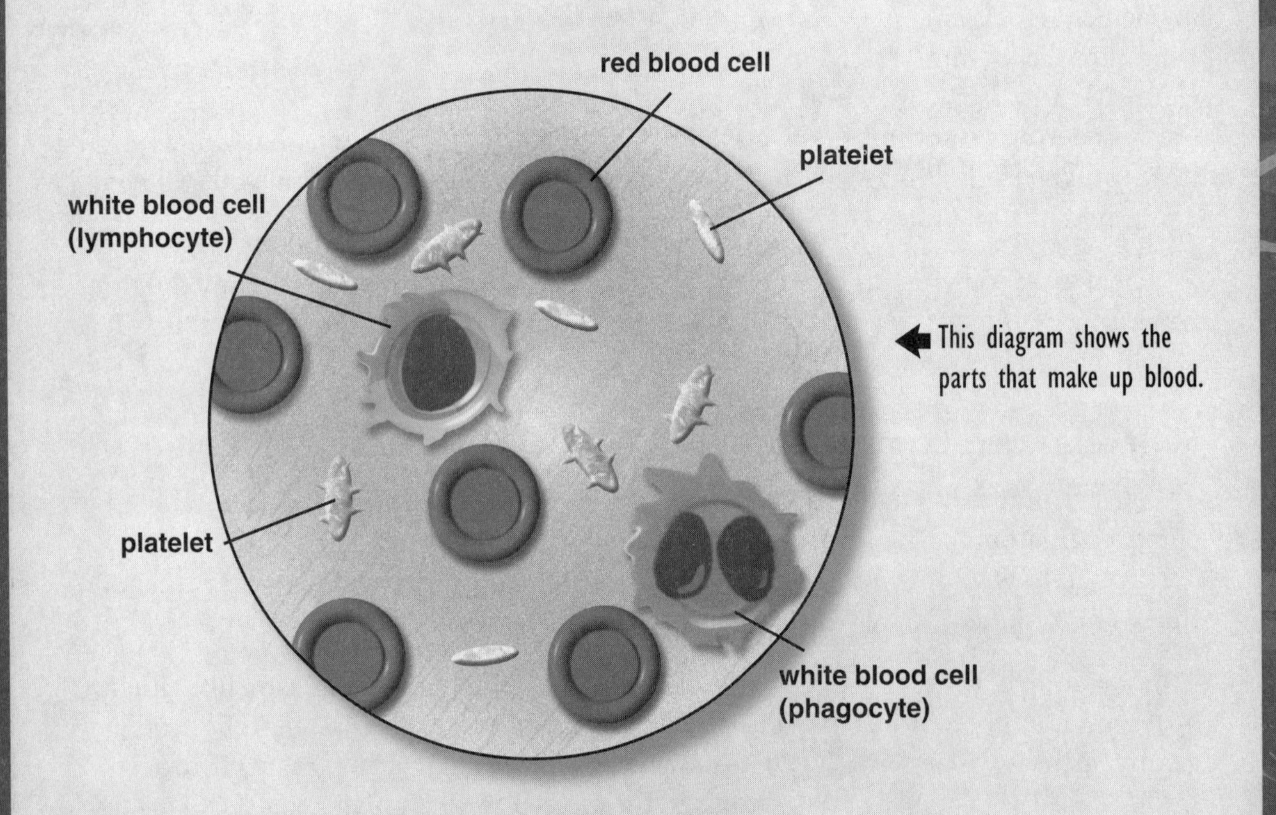

This diagram shows the parts that make up blood.

Comprehension Question

Describe the role of atoms in elements, compounds, and mixtures.

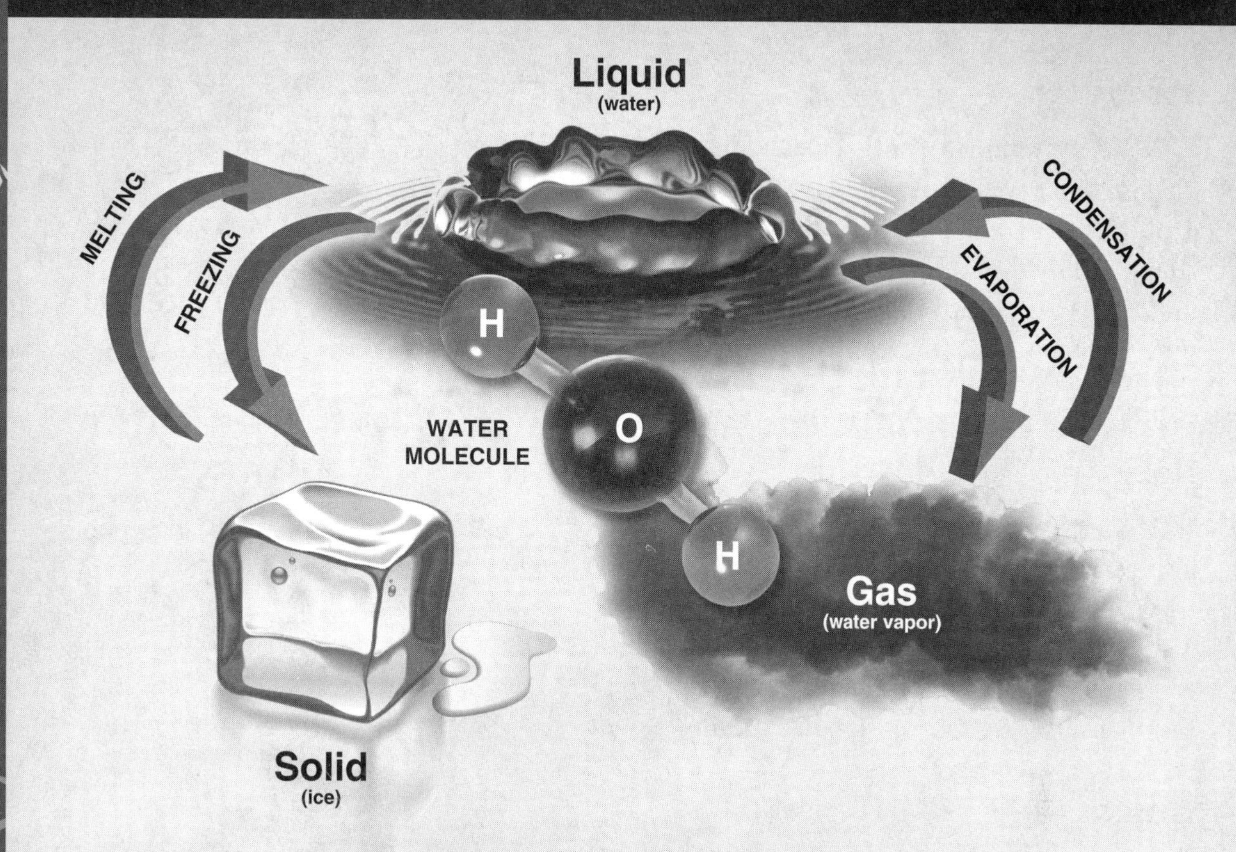

States of Matter

You can skate on solid water. You can put solid water in your drink to make it cold. We call solid water ice.

26

You can swim in liquid water. You can drink it. You can take a shower. You can water plants with it. You can fill your dog's dish with it.

55

When water is a gas, it is called water vapor or steam. It is what clouds are made of. You see it as steam from a tea kettle. You can see it going up from a bowl of hot soup. Water vapor will not stay in your dog's dish. 104

All matter can be a solid. It can be a liquid. It can be a gas. These are the states of matter. 126

Matter can change from one state to another. It may melt. It may evaporate (ee-VAP-uh-rate). It may freeze. However, that never changes the molecules. They stay the same. Water molecules are always the same. They are the same as ice, water, or vapor.

Most things get bigger when they heat up. The molecules shake more. They push each other away. Molecules act one way in solids. They act a different way in liquids. They act one more way in gases. They move most in gases. Then they are the farthest apart. They move less in solids. They are packed close and tight.

Solids

Solids are made of molecules. They are packed close together. They don't move around. They just shake. Solids keep their shape. They don't flow or ooze. Their molecules don't move like that. It is hard to pinch solids. Their molecules are already close together.

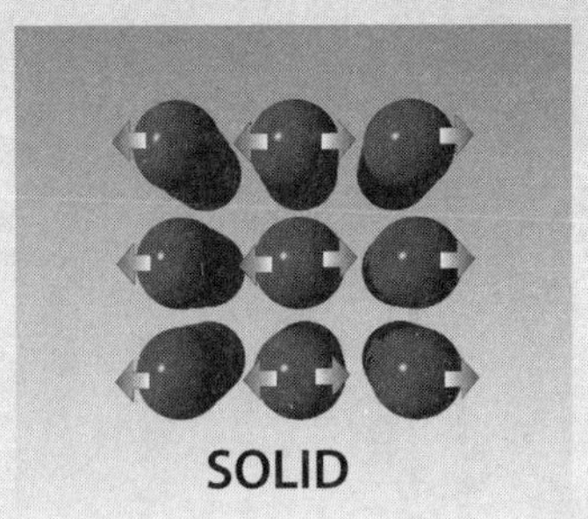

Liquids

Liquids are made of molecules. Then they are farther apart. They move around. Liquids flow. They change shape. They spread out. They make puddles. Liquids fill the bottom of cups and bowls. A liquid always takes up the same amount of space. It won't get any bigger.

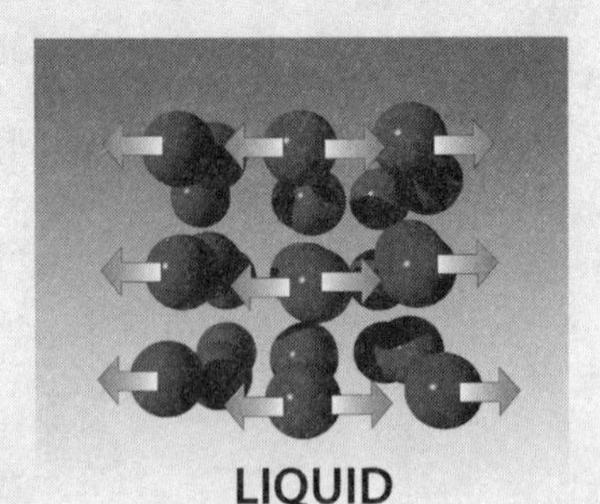

Gases

Gases are made of molecules. Now they are very far apart. They move all over the place. They spread out in all directions. A gas fills all the space in a can. It is easy to squish a gas. Its molecules are far apart. They have lots of space to squeeze into.

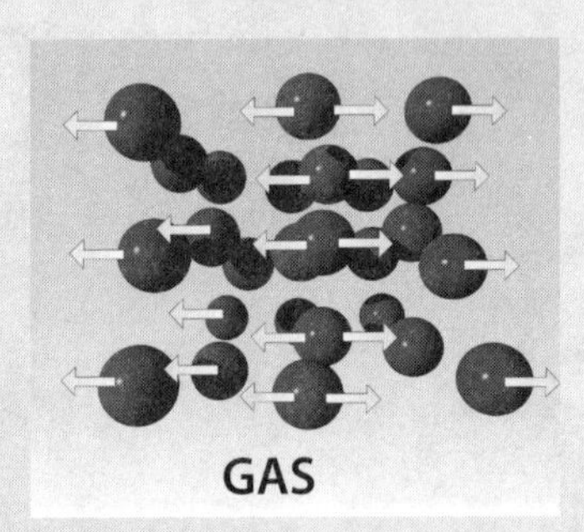

Cool Down!

You can cool things down by taking heat away. It takes energy away. The molecules slow down. You can go down to –273 degrees Celsius (SEL-see-uhs). This is very cold. This is the coldest anything can get. It is called absolute (AB-so-loot) zero. Molecules barely shake. No more energy can be taken away. Everything is a solid.

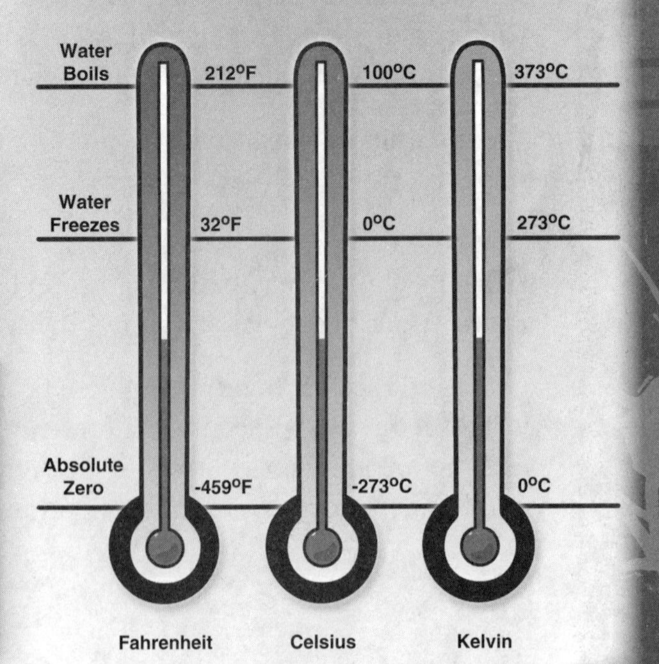

Comprehension Question

How are ice, water, and steam alike? How are they not alike?

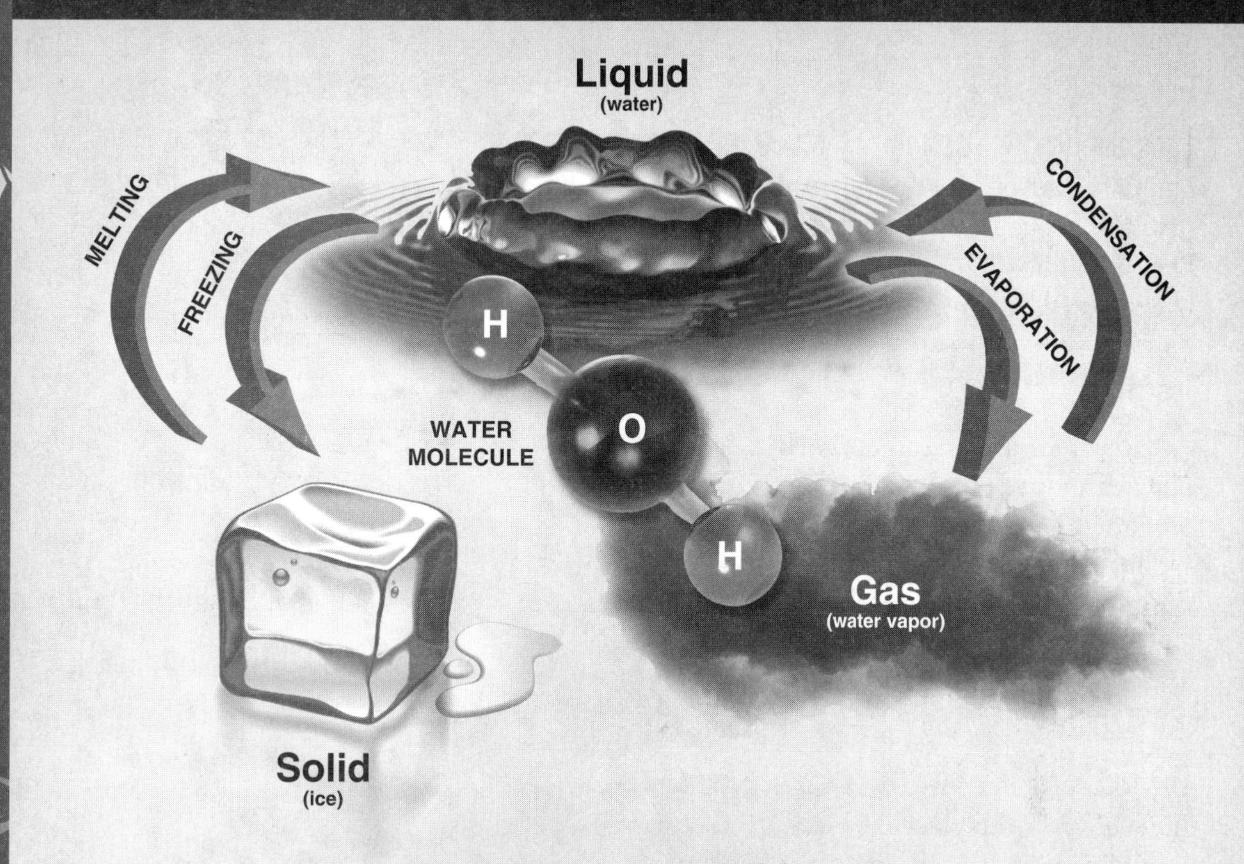

States of Matter

When water is solid, you can skate on it. You can put it in your drinks to make them cold. We call this ice.

When water is a liquid, you can swim in it. You can drink it. You can take a shower. You can water plants with it. You can fill your dog's bowl with it.

When water is a gas, it is called water vapor or steam. This is what clouds are made of. You see it as steam from a tea kettle. You can see it rising off a bowl of hot soup. Water vapor will not stay in your dog's bowl.

All matter can be a solid, a liquid, or a gas. These are the states of matter.

Things can change from one state of matter to another. For example, they may melt or evaporate (ee-VAP-uh-rate). However, changing the state won't change the molecules. They stay the same. Water molecules are the same whether they are ice, water, or vapor.

Most things expand when they heat up. This is because the molecules vibrate more. They push away from each other. Molecules act differently in solids, liquids, and gases. The most active molecules are in gases. They are also the farthest apart. The least active are in solids. They are the closest together.

Solids

Molecules in a solid are packed together. They are packed close. They only move by vibrating. That is why solids keep their shape. They don't flow. It is hard to squish solids. Their molecules are already close together.

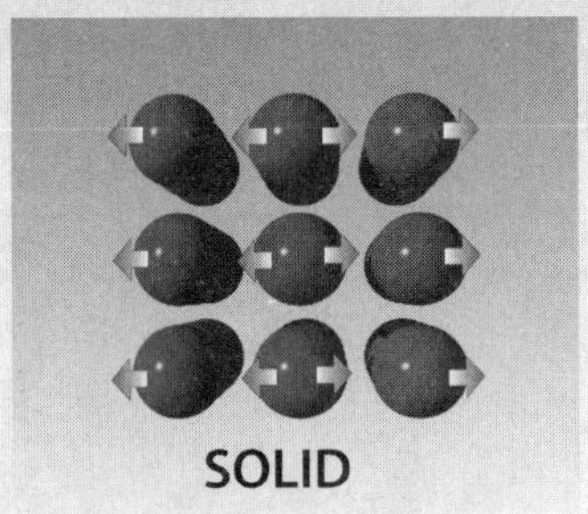

Liquids

Molecules in a liquid are farther apart. They can move past each other. Liquids flow and change shape. They spread out to make puddles. Liquids fill the bottom of any container they are in. It will still take up the same amount of space. It won't get any bigger.

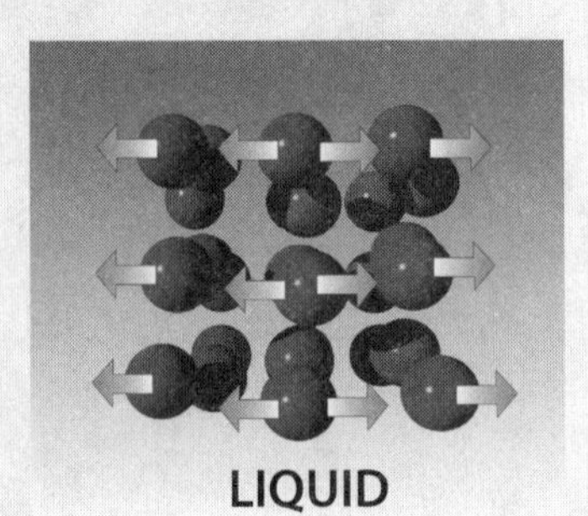

Gases

Molecules in a gas are far apart. They can move all over the place. A gas spreads out in all directions. A gas fills all the space in a container. It is easy to squish a gas because the molecules are far apart.

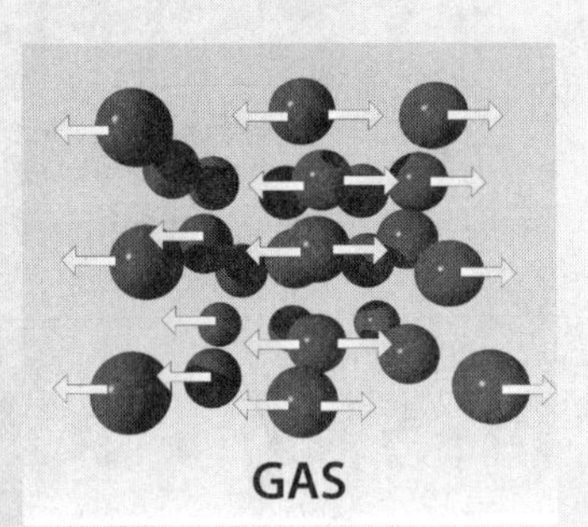

Cool Down!

When you cool something, you take energy from it. The colder it is, the less its atoms move. At minus 273 degrees Celsius (SEL-see-uhs), it is very cold. Molecules vibrate as little as possible then. This is the coldest anything can get. It is called absolute (AB-suh-loot) zero. No more energy can be removed from a thing when it is this cold. It is so cold that everything is a solid.

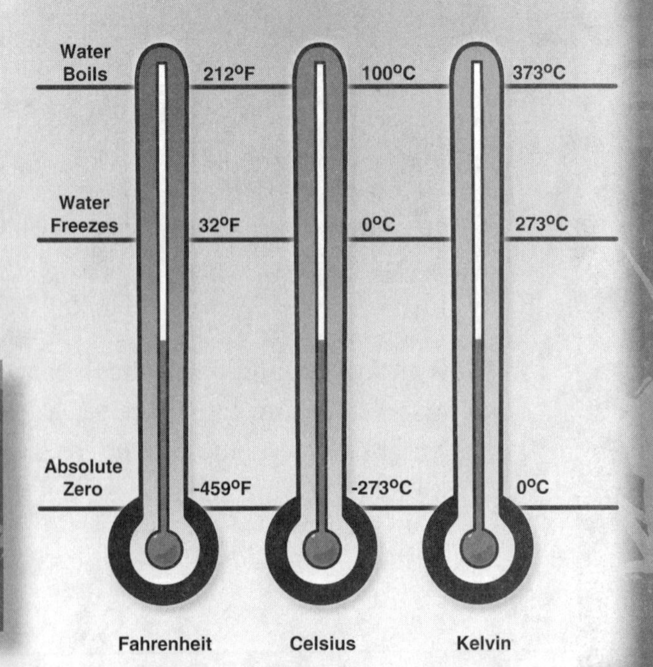

Comprehension Question

Describe the differences between solids, liquids, and gases.

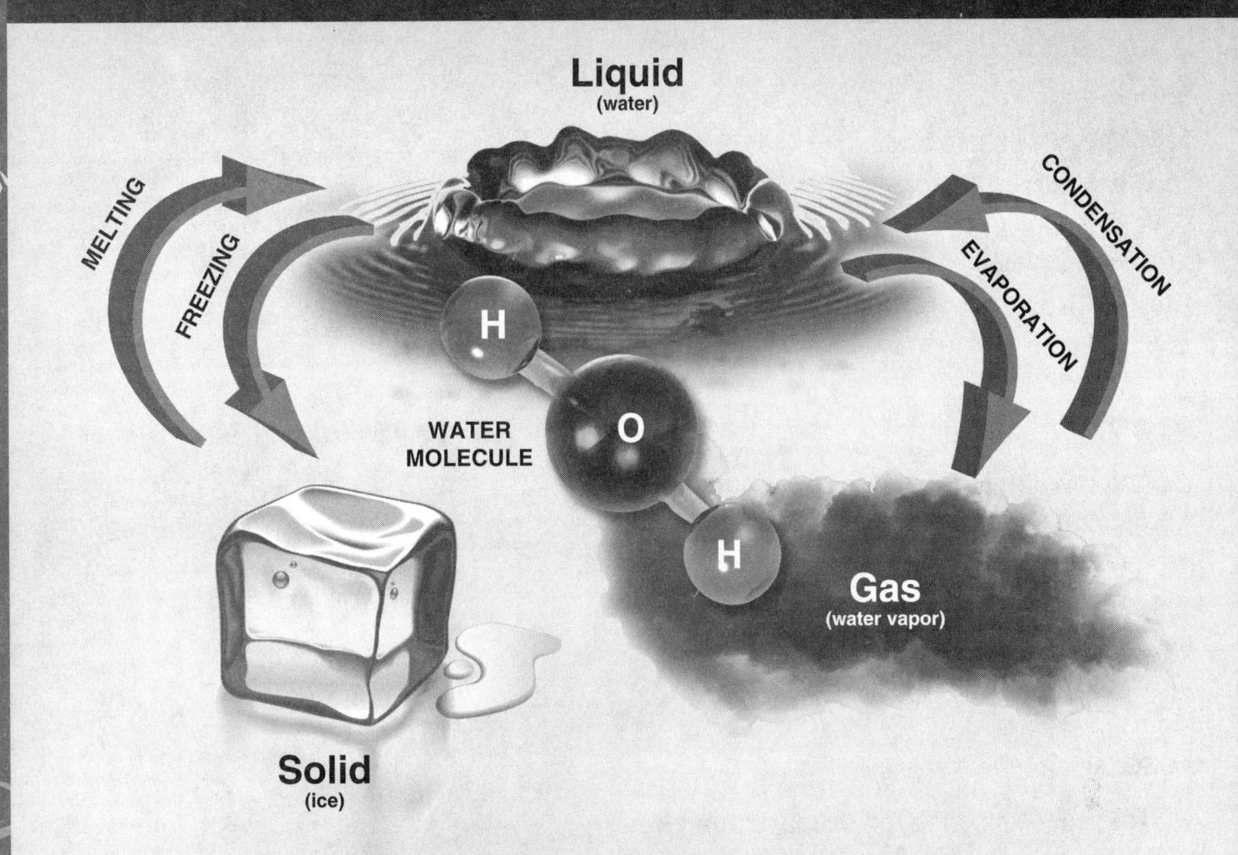

States of Matter

When water is solid, you can skate on it. You can put it in your drinks to make them cold–we call this ice. When water is a liquid, you can swim in it, drink it, take a shower in it, or water plants with it. You can fill your dog's water bowl with it. When water is a gas, it is called water vapor; it is the stuff that clouds are made of. Water vapor would never stay in your dog's water bowl. You see it as steam from a kettle or rising off a bowl of hot soup. 102

There are three states of matter: solid, liquid, and gas. All matter can exist in any and all of these states. 123

Substances change from one state of matter to another at different temperatures. For example, they may melt or evaporate (uh-VAP-uh-rate). However, changing the state won't change the molecules. Water molecules are still water molecules whether they are ice, water, or vapor. 163

Most substances expand when they heat up. This is because the molecules vibrate more and push away from each other. The amount molecules move is different for solids, liquids, and gases. The most active molecules are in gases, where they are the farthest apart. The least active are in solids, where they are packed closest together.

Solids

Molecules in a solid are packed together closely in fixed positions. They can only move by vibrating in these positions. That is why solids keep their shape and don't flow. It is hard to compress solids because their molecules are already close together.

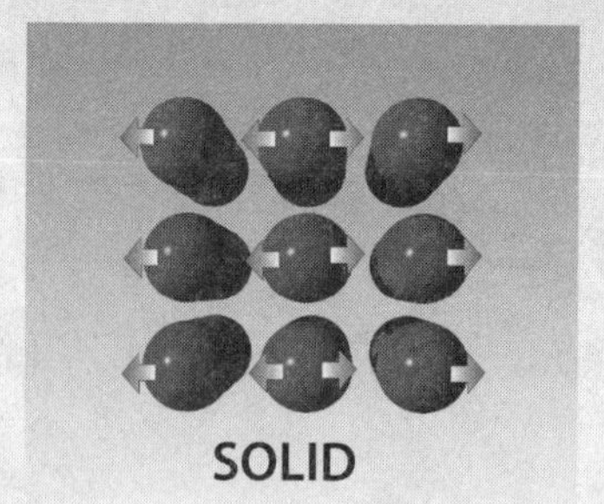

Liquids

Molecules in a liquid are farther apart and can move past each other easily. Liquids can flow and change shape. They can spread out to make puddles. A liquid will fill the bottom of any container it is in. It will still keep the same volume; it won't get any bigger.

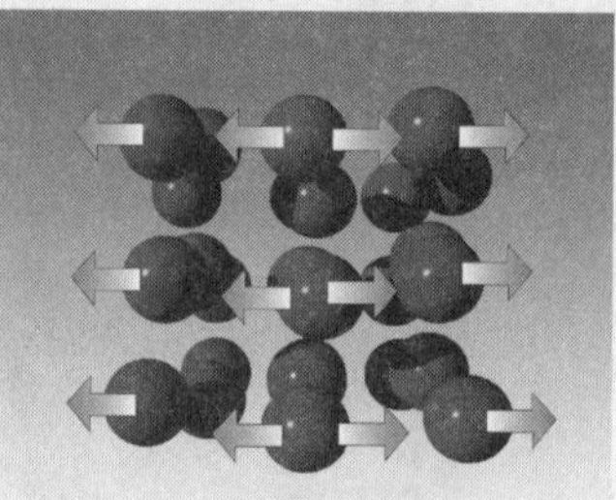

Gases

Molecules in a gas are far apart compared to solids or liquids. They can move freely, and will fill all the space in a container. It is easy to compress a gas because the molecules are far apart.

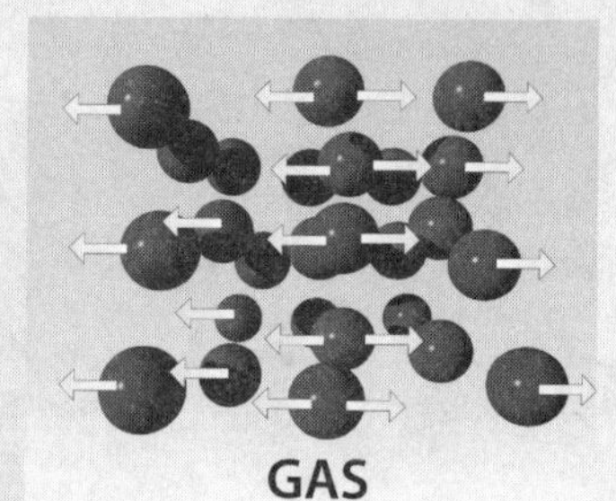

Cool Down!

When you cool something, you take energy from it. The colder it is, the less its atoms move. At minus 273 degrees Celsius (SEL-see-uhs), it is very cold. It is so cold that nearly everything will be a solid. Molecules vibrate as little as possible. This is the coldest anything can get. It is called absolute (AB-suh-loot) zero. No more energy can be removed from something at absolute zero.

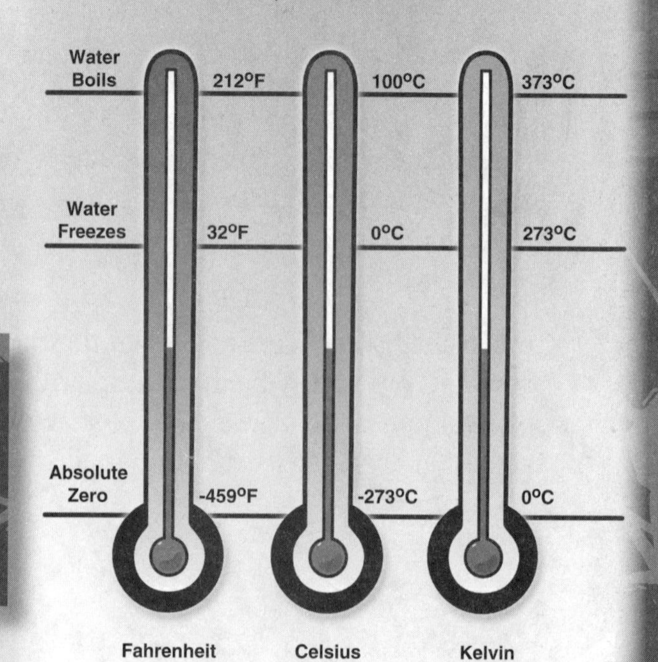

Comprehension Question

Describe how heat determines a substance's state of matter. Use examples.

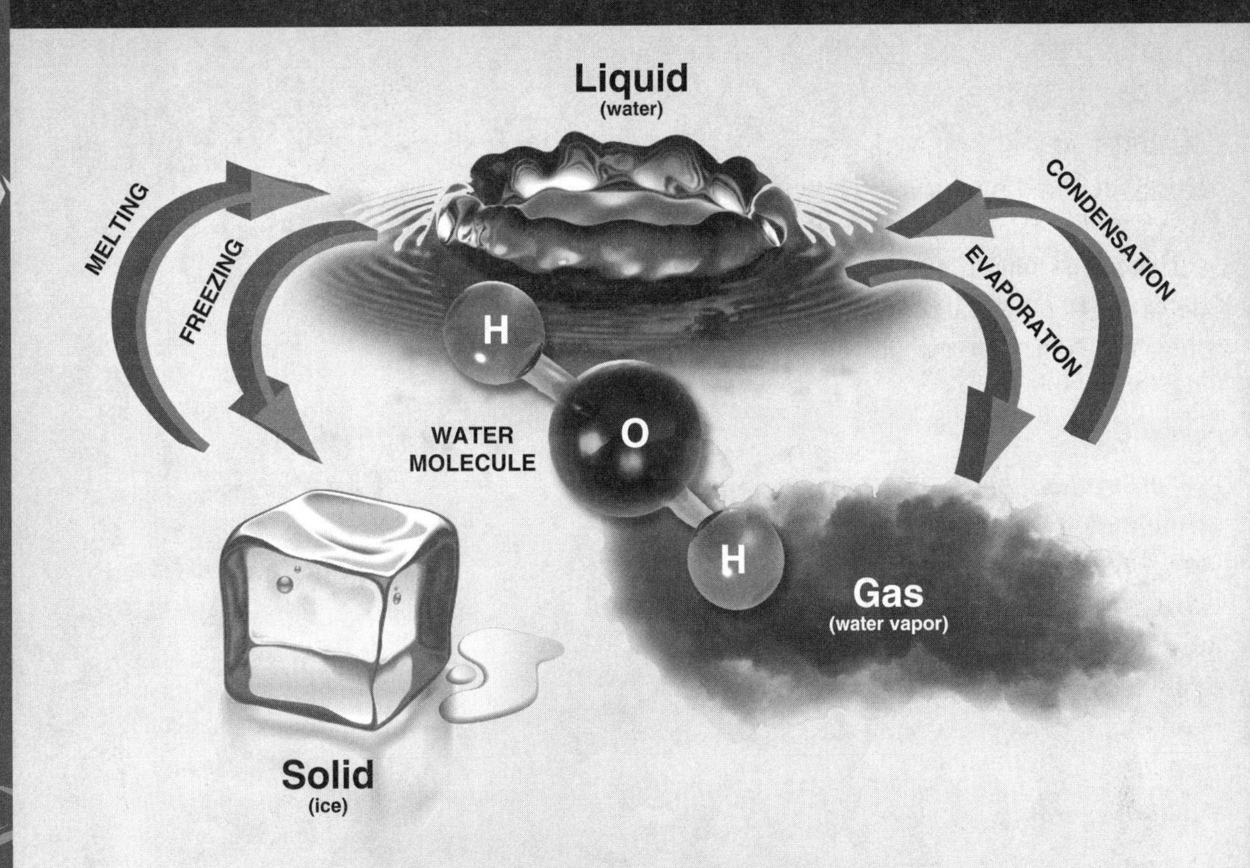

States of Matter

When water is solid, you can skate on it or put it in your drink to make it cold–we call this ice. When water is a liquid, you can drink it, shower in it, water plants with it... even fill your dog's water bowl with it. When water is in its gaseous state, it is water vapor: you see it as steam from a kettle or rising off a bowl of hot soup. Water vapor would never stay in your dog's water bowl. 88

Matter exists in three states: solid, liquid, and gas, and it can change from state to state. 105

Substances change from one state of matter to another at different temperatures. For example, they may melt or evaporate (uh-VAP-uh-rate). However, changing the state never changes the molecules. Water molecules remain water molecules whether they are ice, water, or vapor. 144

Most substances expand when heated. The molecules vibrate more and push away from each other. The amount molecules move is different for solids, liquids, and gases. The most active molecules are in gases, where they are the farthest apart. The least active are in solids, where they are packed closest together.

Solids

Molecules in a solid are packed together closely in fixed positions. They only move by vibrating in these positions. That is why solids keep their shape and don't flow or deform. It is hard to compress solids because their molecules are already close together.

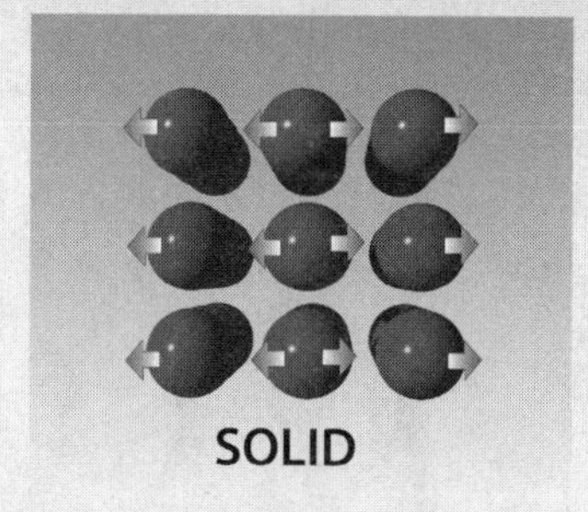

Liquids

Molecules in a liquid are farther apart, which allows them to move around each other easily. Liquids can flow and change shape, spreading out to make puddles or fill the bottom of any container. While its shape may change, it will always keep the same volume; it can't get any bigger.

Gases

Molecules in a gas are very far apart compared to solids or liquids. They move freely and loosely, and because of this they will fill all the space in any container. It is easy to compress a gas because the molecules are far apart.

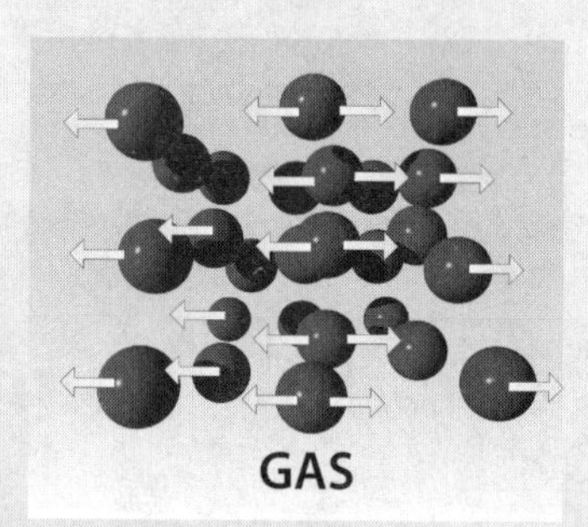

Cool Down!

When you cool something, you take energy from it. The colder it is, the less its atoms move. At minus 273 degrees Celsius (SEL-see-uhs), it is very cold. It is so cold that nearly everything will be a solid. Molecules vibrate as little as possible. This is the coldest anything can get. It is called absolute (AB-suh-loot) zero. No more energy can be removed from something at absolute zero.

Absolute Zero

Thermometers compare Fahrenheit, Celsius, and Kelvin scales.

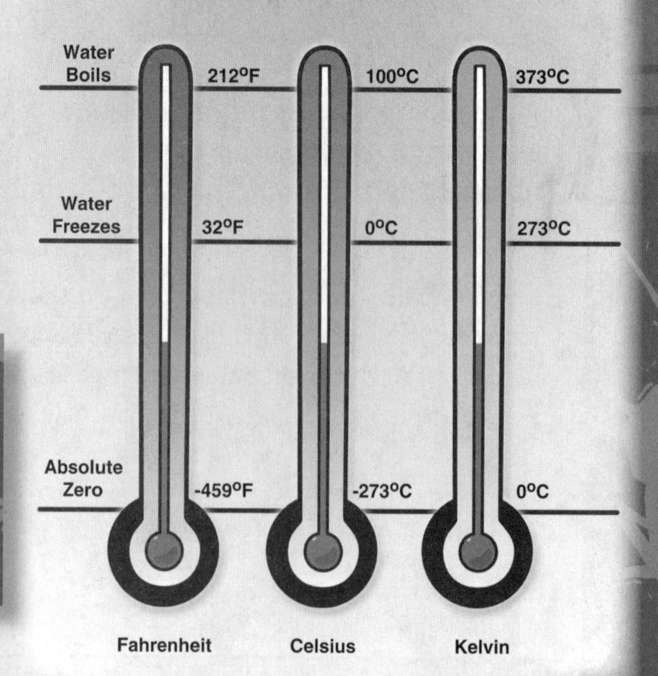

Comprehension Question

Describe the relationship between heat, atoms, and the three states of matter.

The Periodic Table

So Many Elements!

All things are made of atoms. There are more than 100 kinds of atoms. Something can be made with just one kind of atom. That is called an element. Silver is an element. It is made of silver atoms. Helium is also one. It is made of helium atoms.

There are many elements. They all have their own properties (PROP-er-tees). A property can be many things. It can be what an element does. It can be how it acts. It can be how it looks.

Silver atoms share their electrons. This makes them hold each other tight in clumps. Clumps of silver atoms are hard. They keep their shape. They stick together. Some other atoms also share in the same way. They are all called metals.

Helium atoms have lots of electrons. They can't take any more. They do not react with other atoms. A few other kinds of atoms are the same. They also do not react with other atoms. They are called the noble gases.

There are many groups. Some steal electrons. Some react all the time. Others never react. Some fall apart over time. They are called radioactive (RAY-dee-oh-ak-tiv).

Each kind of atom is in lots of groups. Each is in more than one at a time. Uranium is a metal. It is in the metal group. It is also radioactive. It is also in the radioactive group. This can get crazy! It's like all the types of atoms are in a big pile. We needed a way to put them in order.

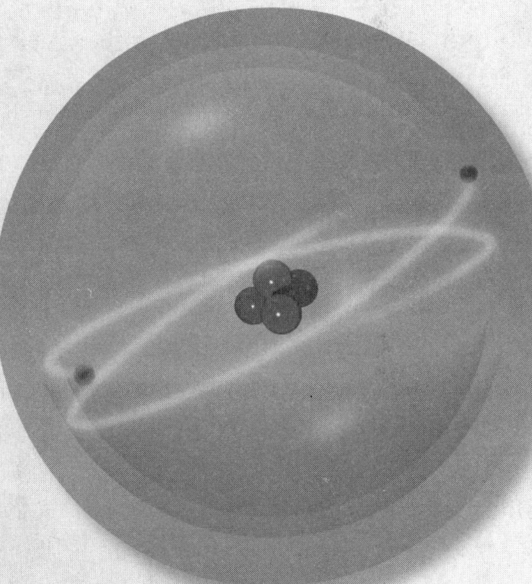

Helium atom

Periodic Table

Dmitri Mendeleév (duh-ME-tree MEN-duh-LAY-yef) was a chemist. He was from the north of Russia. He had lots of brothers. He had even more sisters. He was the youngest. His father died. His mother got him a good education. He loved science. She helped him follow that dream.

Mendeleév learned about the elements. He thought that they should be put in order. He made a chart. It had all the kinds of atoms known. This chart is called the Periodic Table of Elements. It is a tool still used today.

In the table, the atoms are listed in order. They go left to right. They go top to bottom. The order is by atomic number. That is how many protons are in each atom. The table lists the atoms by symbol. Symbols are short forms of their names.

Each row is a period. Each column is a group. Columns can also be called families. All the atoms in a period are the same in some ways. All the atoms in a group are the same in other ways.

When Mendeleév created the table, he put 63 kinds of atoms on it. He thought there were more. He predicted what they would be. He found gaps in his chart. He said the other atoms would fill those gaps. He left space for them. During his life, three new kinds of atoms were discovered. They fit into three holes on the chart. Mendeleév's chart made sense of the pile of elements!

ПЕРИОДИЧЕСКАЯ СИСТЕМА ЭЛЕМЕНТОВ

Периоды	Ряды	I	II	III	IV	V	VI	VII	VIII			0
		ГРУППЫ ЭЛЕМЕНТОВ										
1	I	H 1 1,008										He 2 4,003
2	II	Li 3 6,940	Be 4 9,02	5 B 10,82	6 C 12,010	7 N 14,008	8 O 16,000	9 F 19,00				Ne 10 20,183
3	III	Na 11 22,997	Mg 12 24,32	13 Al 26,97	14 Si 28,06	15 P 30,98	16 S 32,06	17 Cl 35,457				Ar 18 39,944
4	IV	K 19 39,096	Ca 20 40,08	Sc 21 45,10	Ti 22 47,90	V 23 50,95	Cr 24 52,01	Mn 25 54,93	Fe 26 55,85	Co 27 58,94	Ni 28 58,69	
	V	29 Cu 63,57	30 Zn 65,38	31 Ga 69,72	32 Ge 72,60	33 As 74,91	34 Se 78,96	35 Br 79,916				Kr 36 83,7
5	VI	Rb 37 85,48	Sr 38 87,63	Y 39 88,92	Zr 40 91,22	Nb 41 92,91	Mo 42 95,95	Ma 43 —	Ru 44 101,7	Rh 45 102,91	Pd 46 106,7	
	VII	47 Ag 107,88	48 Cd 112,41	49 In 114,76	50 Sn 118,70	51 Sb 121,76	52 Te 127,61	53 J 126,92				Xe 54 131,3
6	VIII	Cs 55 132,91	Ba 56 137,36	La 57 * 138,92	Hf 72 178,6	Ta 73 180,88	W 74 183,92	Re 75 186,31	Os 76 190,2	Ir 77 193,1	Pt 78 195,23	
	IX	79 Au 197,2	80 Hg 200,61	81 Tl 204,39	82 Pb 207,21	83 Bi 209,00	84 Po 210	85 —				Rn 86 222
7	X	87 —	Ra 88 226,05	Ac 89 227	Th 90 232,12	Pa 91 231	U 92 238,07					

* ЛАНТАНИДЫ 58-71

Ce 58 140,13	Pr 59 140,92	Nd 60 144,27	61 —	Sm 62 150,43	Eu 63 152,0	Gd 64 156,9
Tb 65 [illegible]	Dy 66 162,46	Ho 67 164,94	Er 68 167,2	Tu 69 169,4	Yb 70 173,04	Cp 71 174,99

Comprehension Question

Why did Mendeleév make his chart?

The Periodic Table

So Many Elements!

All things are made of atoms. There are more than 100 kinds of atoms. Something can be made with just one kind of atom. That is called an element. Silver is an element. It is made of silver atoms. Helium is also an element. It is made of helium atoms.

There are many different elements. They all have different properties (PROP-er-tees). A property can be many things. It can be what an element does. It can be how it acts. It can be how it looks.

Silver atoms share their electrons. This makes them hold each other tight in clumps. Clumps of silver atoms are hard. They keep their shape. They stick together. Some other elements also share their electrons. They are all called metals.

Helium atoms have lots of electrons. They can't take any more. They do not react with other atoms. A few other elements are the same. They have lots of electrons. They also do not react with other atoms. They are called the noble gases.

There are many groups. Some steal electrons. Some react all the time. Others never react. Some fall apart over time. They are called radioactive (RAY-dee-oh-ak-tiv).

Elements are in lots of groups. They are in more than one at a time. Uranium is a metal. It is also radioactive. It is in the metal group. It is also in the radioactive group. This can be very confusing. It's like all the elements are in a big pile. Scientists needed a way to organize them.

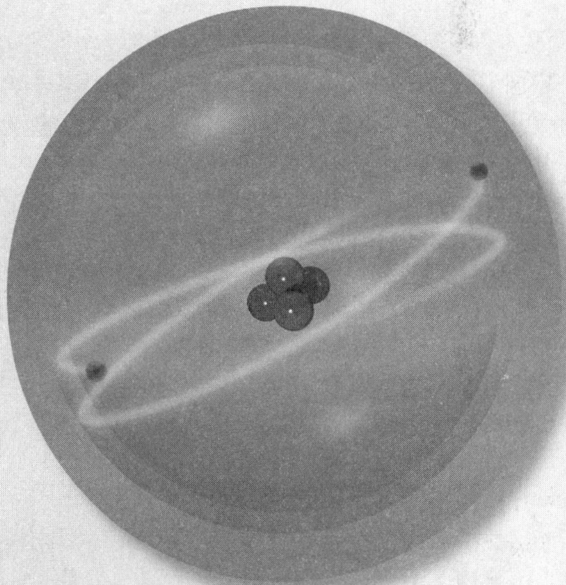

Helium atom

Periodic Table

Dmitri Mendeleév (duh-ME-tree MEN-duh-LAY-yef) was a chemist. He was from Siberia. He had lots of brothers and sisters. He was the youngest. His father died. His mother made sure he got a good education. He loved science. She helped him follow that dream.

Mendeleév learned about the elements. He thought they should be organized. He made a chart. It had all the known elements. This chart is called the Periodic Table of Elements. It is a tool still used by scientists today.

In the table, elements are listed in order. They go left to right and top to bottom. The order is by atomic number. That is the number of protons in each atom. The table lists the elements by symbol. These are short forms of their names.

Each row in the chart is called a period. Each column is called a group. Columns can also be called families. All the elements in a period share some properties. All the elements in a group share other properties.

When Mendeleév created the table, there were 63 known elements. He believed there were more. He predicted what there would be. He found gaps in his chart. He said the elements would fill those gaps. He left space for them on purpose. During his life, three of the elements were discovered. They fit into three holes on the chart. Mendeleév's chart made sense of the pile of elements!

ПЕРИОДИЧЕСКАЯ СИСТЕМА ЭЛЕМЕНТОВ

ГРУППЫ ЭЛЕМЕНТОВ

		I	II	III	IV	V	VI	VII	VIII			0
1	I	H 1 1,008										He 2 4,003
2	II	Li 3 6,940	Be 4 9,02	5 B 10,82	6 C 12,010	7 N 14,008	8 O 16,000	9 F 19,00				Ne 10 20,183
3	III	Na 11 22,997	Mg 12 24,32	13 Al 26,97	14 Si 28,06	15 P 30,98	16 S 32,06	17 Cl 35,457				Ar 18 39,944
4	IV	K 19 39,096	Ca 20 40,08	Sc 21 45,10	Ti 22 47,90	V 23 50,95	Cr 24 52,01	Mn 25 54,93	Fe 26 55,85	Co 27 58,94	Ni 28 58,69	
	V	29 Cu 63,57	30 Zn 65,38	31 Ga 69,72	32 Ge 72,60	33 As 74,91	34 Se 78,96	35 Br 79,916				Kr 36 83,7
5	VI	Rb 37 85,48	Sr 38 87,63	Y 39 88,92	Zr 40 91,22	Nb 41 92,91	Mo 42 95,95	Ma 43 —	Ru 44 101,7	Rh 45 102,91	Pd 46 106,7	
	VII	47 Ag 107,88	48 Cd 112,41	49 In 114,76	50 Sn 118,70	51 Sb 121,76	52 Te 127,61	53 J 126,92				Xe 54 131,3
6	VIII	Cs 55 132,91	Ba 56 137,36	La 57 * 138,92	Hf 72 178,6	Ta 73 180,88	W 74 183,92	Re 75 186,31	Os 76 190,2	Ir 77 193,1	Pt 78 195,23	
	IX	79 Au 197,2	80 Hg 200,61	81 Tl 204,39	82 Pb 207,21	83 Bi 209,00	84 Po 210	85 —				Rn 86 222
7	X	— 87	Ra 88 226,05	Ac 89 227	Th 90 232,12	Pa 91 231	U 92 238,07					

* ЛАНТАНИДЫ 58–71

Ce 58 140,13	Pr 59 140,92	Nd 60 144,27	61 —	Sm 62 150,43	Eu 63 152,0	Gd 64 156,9
Tb 65 [illegible]	Dy 66 162,46	Ho 67 164,94	Er 68 167,2	Tu 69 169,4	Yb 70 173,04	Cp 71 174,99

Comprehension Question

Why is the periodic table useful to scientists?

The Periodic Table

So Many Elements!

Everything is made of atoms. There are over 100 different kinds of atoms. When something is made entirely of one kind of atom, it is called an element. Silver is an element. It is made of silver atoms. Helium is another element. It is made of helium atoms.

Different elements have different properties (PROP-er-tees). A property can be something that an element does. It can be how it acts. It can be how it looks.

For instance, silver atoms share their electrons in a special way. This makes them easy to pack together. Clumps of silver atoms are hard. They keep their shape because they pack together. A few other elements also share electrons in this way. Together, these elements are called metals.

Helium atoms have so many electrons that they can't take any more. That means that they do not react with other atoms. The atoms of a few other elements are also full up on electrons. They do not react with other atoms, either. This group of elements is called the noble gases.

Helium atom

There are many different groups of elements. There are some elements that steal electrons from other elements. There are some elements that react in one way, and other elements that react the other way. There are some elements that fall apart over time. They are called radioactive elements.

Elements are members of more than one group at a time. Uranium is a metal. It is also radioactive. It is in the metal group. It is also in the radioactive group. This can be very confusing. It's like there is just a big pile of elements. Scientists didn't know how to organize all of these elements until Dmitri Mendeleév (duh-ME-tree MEN-duh-LAY-yef).

Periodic Table

Dmitri Mendeleév was a chemist from Siberia. He was the youngest of 14 children. His father died. His mother made sure her young son got a good education. He loved science. She helped him follow that dream.

Mendeleév studied chemistry. He learned about the elements. He thought they should be organized. In 1869, he decided to organize all the known elements into a chart. This chart is called the Periodic Table of Elements. It is a tool still used by scientists today.

In the table, elements are listed left to right and top to bottom. They are listed by their atomic number. The atomic number is the number of protons in each atom. The table lists the elements' chemical symbols. The symbols are short forms of their names.

The rows of elements are called periods. Each column of the table is called a group or family. The elements in these periods and groups share properties.

When Mendeleév created the table, there were 63 known elements. He believed there were more. He also thought the unknown elements could be predicted. He found gaps in his table. He believed that elements would be found to fill those gaps. He left space for them on purpose. During his life, three of the elements he predicted were discovered. Mendeleév made sense of the pile of elements!

ПЕРИОДИЧЕСКАЯ СИСТЕМА ЭЛЕМЕНТОВ

		ГРУППЫ ЭЛЕМЕНТОВ										
		I	II	III	IV	V	VI	VII	VIII			0
1	I	H 1 1,008										He 2 4,003
2	II	Li 3 6,940	Be 4 9,02	5 B 10,82	6 C 12,010	7 N 14,008	8 O 16,000	9 F 19,00				Ne 10 20,183
3	III	Na 11 22,997	Mg 12 24,32	13 Al 26,97	14 Si 28,06	15 P 30,98	16 S 32,06	17 Cl 35,457				Ar 18 39,944
4	IV	K 19 39,096	Ca 20 40,08	Sc 21 45,10	Ti 22 47,90	V 23 50,95	Cr 24 52,01	Mn 25 54,93	Fe 26 55,85	Co 27 58,94	Ni 28 58,69	
	V	29 Cu 63,57	30 Zn 65,38	31 Ga 69,72	32 Ge 72,60	33 As 74,91	34 Se 78,96	35 Br 79,916				Kr 36 83,7
5	VI	Rb 37 85,48	Sr 38 87,63	Y 39 88,92	Zr 40 91,22	Nb 41 92,91	Mo 42 95,95	Ma 43 —	Ru 44 101,7	Rh 45 102,91	Pd 46 106,7	
	VII	47 Ag 107,88	48 Cd 112,41	49 In 114,76	50 Sn 118,70	51 Sb 121,76	52 Te 127,61	53 J 126,92				Xe 54 131,3
6	VIII	Cs 55 132,91	Ba 56 137,36	La 57 * 138,92	Hf 72 178,6	Ta 73 180,88	W 74 183,92	Re 75 186,31	Os 76 190,2	Ir 77 193,1	Pt 78 195,23	
	IX	79 Au 197,2	80 Hg 200,61	81 Tl 204,39	82 Pb 207,21	83 Bi 209,00	84 Po 210	85 —				Rn 86 222
7	X	87 —	Ra 88 226,05	Ac 89 227	Th 90 232,12	Pa 91 231	U 92 238,07					

* ЛАНТАНИДЫ 58–71

Ce 58 140,13	Pr 59 140,92	Nd 60 144,27	61 —	Sm 62 150,43	Eu 63 152,0	Gd 64 156,9
Tb 65 [illegible]	Dy 66 162,46	Ho 67 164,94	Er 68 167,2	Tu 69 169,4	Yb 70 173,04	Cp 71 174,99

Comprehension Question

How does the periodic table present the properties of different elements?

The Periodic Table

So Many Elements!

Everything in the world is made of atoms, and there are over 100 different kinds of atoms. When something is made entirely of one kind of atom, it is called an element. Silver is an element, made of silver atoms. Helium is another element, made of helium atoms.

Different elements have different properties (PROP-er-tees). A property is something that an element does or how it acts in certain circumstances.

For instance, silver atoms share their electrons in a special way. This makes them easy to pack together. Clumps of silver atoms are hard and keep their shape because they pack together in this way. A few other elements also share electrons in this special way. Together, these elements are called metals.

Helium atoms have so many electrons that they can't take any more. That means that they do not react with other atoms. A few other atoms are also full up on electrons. They do not react with other atoms, either. This group of elements is called the noble gases.

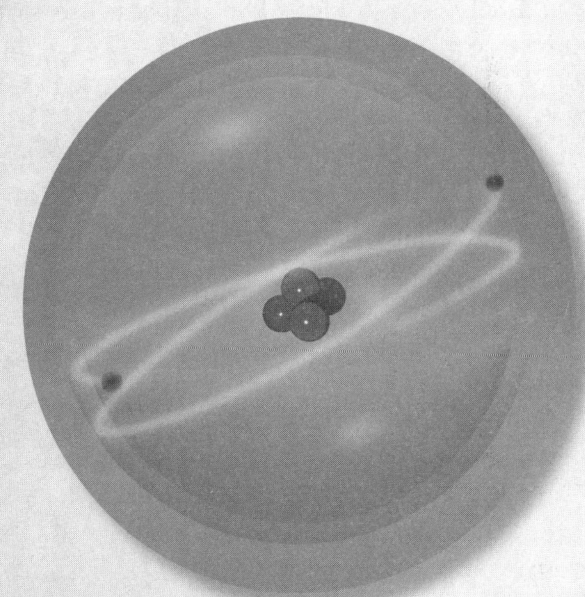

Helium atom

There are many different groups of elements. There are some elements that react in one way, and other elements that react in another way. There are some elements that fall apart over time called radioactive elements. There are some elements that steal electrons from other elements.

Elements are members of more than one group at a time. Uranium, for instance, is a metal and is also radioactive. It is in the metal group and in the radioactive group. This can be very confusing. It's like there is just a big pile of elements. Scientists didn't know how to organize all of these elements until Dmitri Mendeleév (duh-ME-tree MEN-duh-LAY-yef).

Periodic Table

Dmitri Mendeleév was a chemist from Siberia. He was the youngest of 14 children. When his father died, his mother put all of her energy into getting her young son a good education. He loved science, so she helped him follow that dream.

Mendeleév studied chemistry. He learned about the elements, but he thought they should be organized. In 1869, he decided to organize all the known elements into a chart. This chart is called the Periodic Table of Elements. It remains a basic tool used by scientists today.

In the table, elements are arranged left to right and top to bottom. They are arranged by their atomic number. The atomic number is the number of protons in an element's atom. The elements are written in the form of their chemical symbols. The symbols are short forms of the names of the elements.

The rows of elements are called periods. Each column of the table is called a group or family. Elements in these periods and groups share properties.

When Mendeleév created the table, there were 63 known elements. He believed there were more. He also thought the unknown elements could be predicted. He found gaps in his table. He believed that elements would be found to fill those gaps. He left space for them on purpose. During his life, three of the elements he predicted were discovered. Mendeleév made sense of the pile of elements!

ПЕРИОДИЧЕСКАЯ СИСТЕМА ЭЛЕМЕНТОВ

ПЕРИОДЫ	РЯДЫ	ГРУППЫ ЭЛЕМЕНТОВ										
		I	II	III	IV	V	VI	VII	VIII			0
1	I	H 1 1,008										He 2 4,003
2	II	Li 3 6,940	Be 4 9,02	5 B 10,82	6 C 12,010	7 N 14,008	8 O 16,000	9 F 19,00				Ne 10 20,183
3	III	Na 11 22,997	Mg 12 24,32	13 Al 26,97	14 Si 28,06	15 P 30,98	16 S 32,06	17 Cl 35,457				Ar 18 39,944
4	IV	K 19 39,096	Ca 20 40,08	Sc 21 45,10	Ti 22 47,90	V 23 50,95	Cr 24 52,01	Mn 25 54,93	Fe 26 55,85	Co 27 58,94	Ni 28 58,69	
	V	29 Cu 63,57	30 Zn 65,38	31 Ga 69,72	32 Ge 72,60	33 As 74,91	34 Se 78,96	35 Br 79,916				Kr 36 83,7
5	VI	Rb 37 85,48	Sr 38 87,63	Y 39 88,92	Zr 40 91,22	Nb 41 92,91	Mo 42 95,95	Ma 43 —	Ru 44 101,7	Rh 45 102,91	Pd 46 106,7	
	VII	47 Ag 107,88	48 Cd 112,41	49 In 114,76	50 Sn 118,70	51 Sb 121,76	52 Te 127,61	53 J 126,92				Xe 54 131,3
6	VIII	Cs 55 132,91	Ba 56 137,36	La 57 * 138,92	Hf 72 178,6	Ta 73 180,88	W 74 183,92	Re 75 186,31	Os 76 190,2	Ir 77 193,1	Pt 78 195,23	
	IX	79 Au 197,2	80 Hg 200,61	81 Tl 204,39	82 Pb 207,21	83 Bi 209,00	84 Po 210	85 —				Rn 86 222
7	X	87 —	Ra 88 226,05	Ac 89 227	Th 90 232,12	Pa 91 231	U 92 238,07					

* ЛАНТАНИДЫ 58-71

Ce 58 140,13	Pr 59 140,92	Nd 60 144,27	61 —	Sm 62 150,43	Eu 63 152,0	Gd 64 156,9
Tb 65	Dy 66	Ho 67 164,94	Er 68 167,2	Tu 69 169,4	Yb 70 173,04	Cp 71 174,99

Comprehension Question

How is the organization of the periodic table useful for studying the properties of different elements?

Chemical Reactions

Lavoisier

Antoine Lavoisier (ahn-TWAN la-vwah-ZYAY) was a man who studied science a long time ago. He did important work. Now we say that he is the "Father of Modern Chemistry." He changed science for all time.

The Element Oxygen

Lavoisier was the first person to name oxygen. He called it an element. He said an element is a thing that does not break into things that are more simple. But it can be a part of other things. Oxygen is a part of other things. It is a part of air. It is in water, too.

All things are made of atoms. Atoms are small bits of matter. There are many kinds of atoms. An element is made of atoms that are the same.

Atoms are made of three parts. One part is protons. One part is neutrons. One part is electrons. Atoms of an element have the same parts. They are mixed the same way. Parts mixed in other ways make other kinds of atoms. Other kinds of atoms make other elements. Lavoisier named 34 elements.

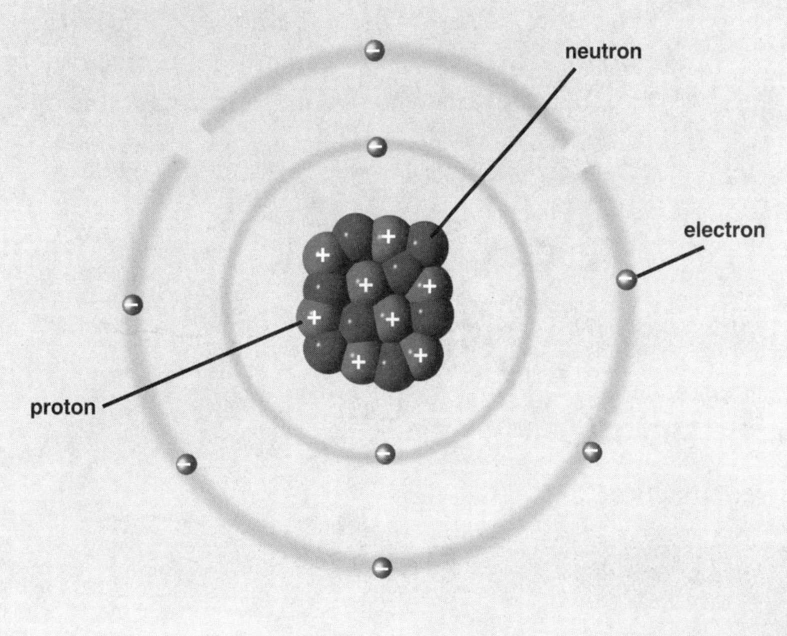

Oxygen atom

Law of Conservation of Mass

Two substances can have a chemical reaction. That means they react with each other. They change. They become a new substance. Oxygen and iron are two substances. They can react with each other. They change when they react. They become rust.

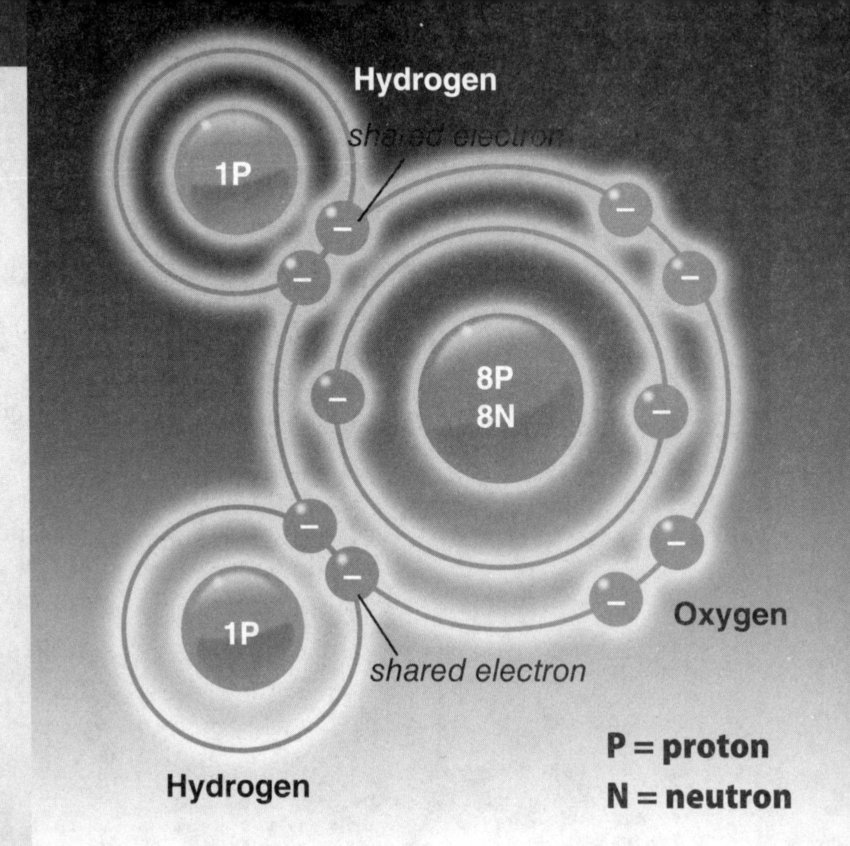

Lavoisier wanted to know what happens. He made a plan. He measured the mass of each substance first. Mass is the amount of matter in the substance. Next, he let the substances react with each other. They made a new substance. He measured the mass of the new substance. The mass before was the same as the mass after!

Oxidation

Lavoisier learned one more thing about oxygen. It reacts with many things. But it often reacts in the same way. It rusts iron. It makes copper go dull. It turns some metals gray. There is one reason for this. That reason is oxidation (OX-id-ay-shun). Lavoisier knew this was so. But he did not live to find out more.

We know more now. We know that oxygen atoms have a "hole." Electrons are on the outside of the atom. That is where the hole is. The hole sucks in electrons from other atoms. This lets the oxygen atom drag the other atoms. When the atoms get stuck together, they make new compounds. Rust is one of these compounds.

Comprehension Question

What is a chemical reaction?

Chemical Reactions

Lavoisier

Antoine Lavoisier (ahn-TWAN la-vwah-ZYAY) was a scientist who lived a long time ago. He is known as the "Founder of Modern Chemistry." His work changed the study of science for all time.

The Element Oxygen

Lavoisier was the first scientist to name oxygen. He said it was an element. An element is a thing that can not break into simpler things. Oxygen is important. It is in our air. It is part of water, too.

We know more about elements now. We know that all things are made of atoms. Atoms are small bits of matter. There are different kinds of atoms. An element is made of atoms that are the same.

All atoms are made of three parts. One part is protons. One part is neutrons. The third part is electrons. The atoms in an element are made of the same mix of these parts. A different mix makes different kinds of atoms. Different kinds of atoms make different elements.

There are over 100 kinds of elements. Lavoisier named 34 of them.

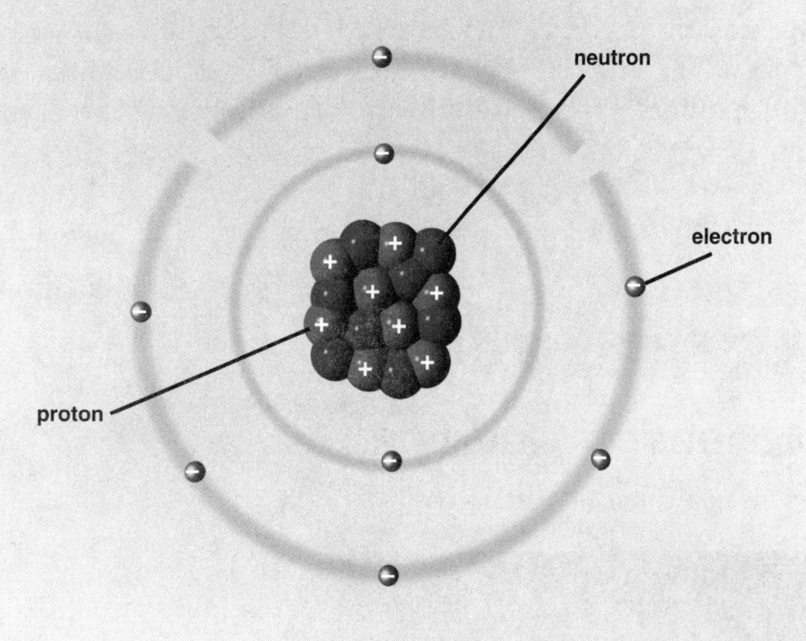

Oxygen atom

Law of Conservation of Mass

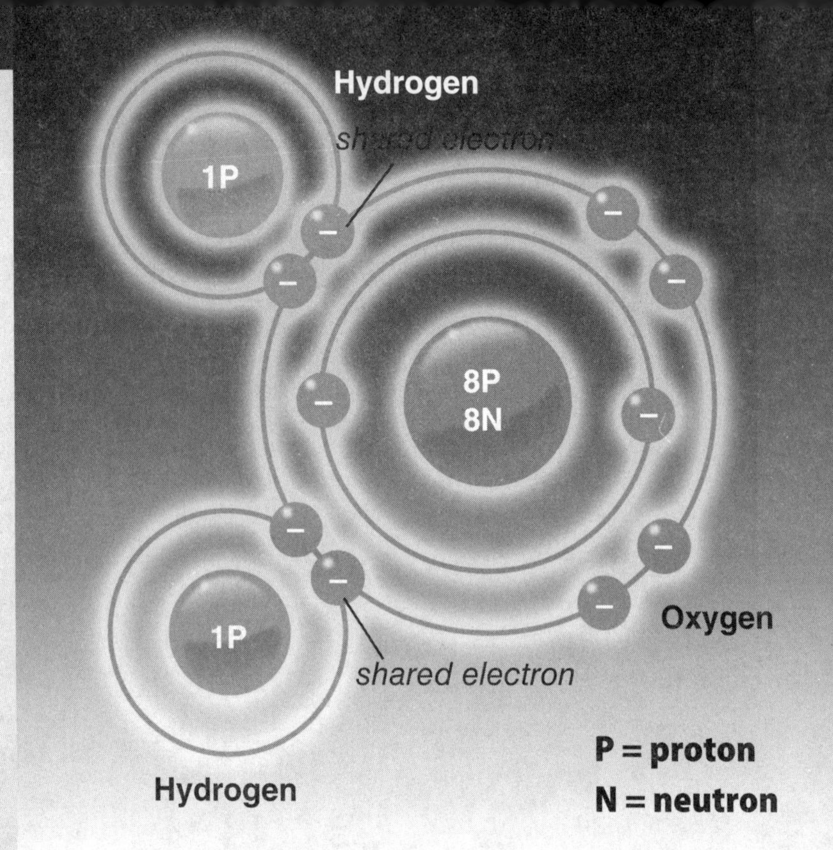

Lavoisier worked on chemical reactions. A chemical reaction happens between substances. They react with each other. They become a new substance. Oxygen and iron react with each other. They become rust.

First, Lavoisier measured the mass of each substance. Mass is the amount of matter a substance has. Then, he let the substances react. He measured the mass of the new substance. The masses were the same! He saw that matter is never lost or gained. The total mass stays the same when something new is made. He called this law Conservation (kon-ser-VAY-shuhn) of Mass. To conserve means to keep it the same. Matter can change. But its mass stays the same.

Oxidation

Lavoisier learned one more thing about oxygen. It reacts with many things. It often reacts in the same way. Oxygen rusts iron. It dulls copper. It turns some metals gray. Why does it do this? Lavoisier said it is because of oxidation (OX-id-ay-shun). Sadly, he died before he found out what it is.

Now we know that oxygen atoms have a "hole." The hole is in the electrons on the outside of the atom. The hole sucks in electrons from other atoms. Then the oxygen drags the other atoms. When the atoms are stuck together, they make new compounds. Rust is one of these compounds.

Comprehension Question

What happens in a chemical reaction?

Chemical Reactions

Lavoisier

Antoine Lavoisier (ahn-TWAN la-vwah-ZYAY) is called the "Founder of Modern Chemistry." He did important work as a scientist. He changed chemistry forever.

The Element Oxygen

Lavoisier was the first to name oxygen. He said it was an element. He defined an element as a thing that could not break down into simpler things. Oxygen is an important element. It is in the air we breathe. It is a part of water, too.

Scientists today have learned more about elements. They know that everything is made of atoms. Atoms are very small pieces of matter. There are many different kinds of atoms. An element is made of the same kind of atoms. All the atoms of an element are made of the same mix of three particles. The particles are protons, neutrons, and electrons. The different mix of particles makes different kinds of atoms. Different kinds of atoms make different elements.

Lavoisier also named 33 other elements! There are over 100 different elements in all. Many were not known until recently.

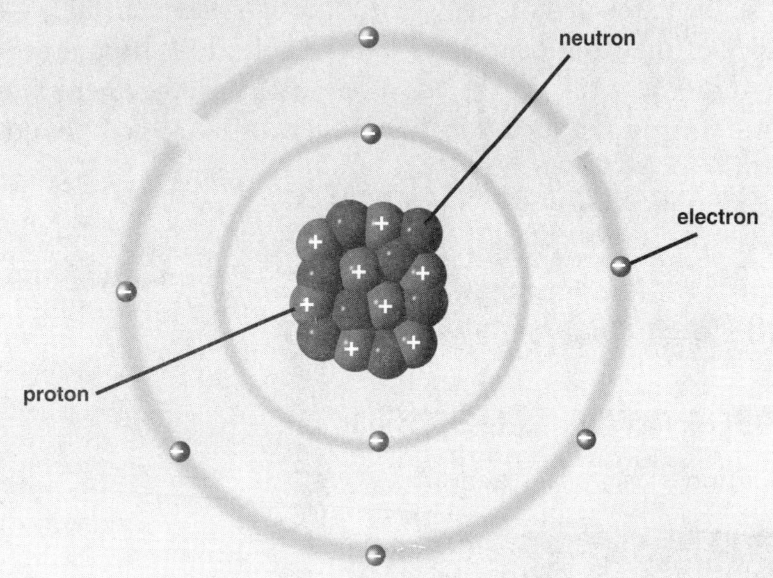

Oxygen atom

Law of Conservation of Mass

Lavoisier worked on chemical reactions. A chemical reaction happens when substances react. They turn into a new substance. When oxygen gas and iron metal come together, they react. They become rust.

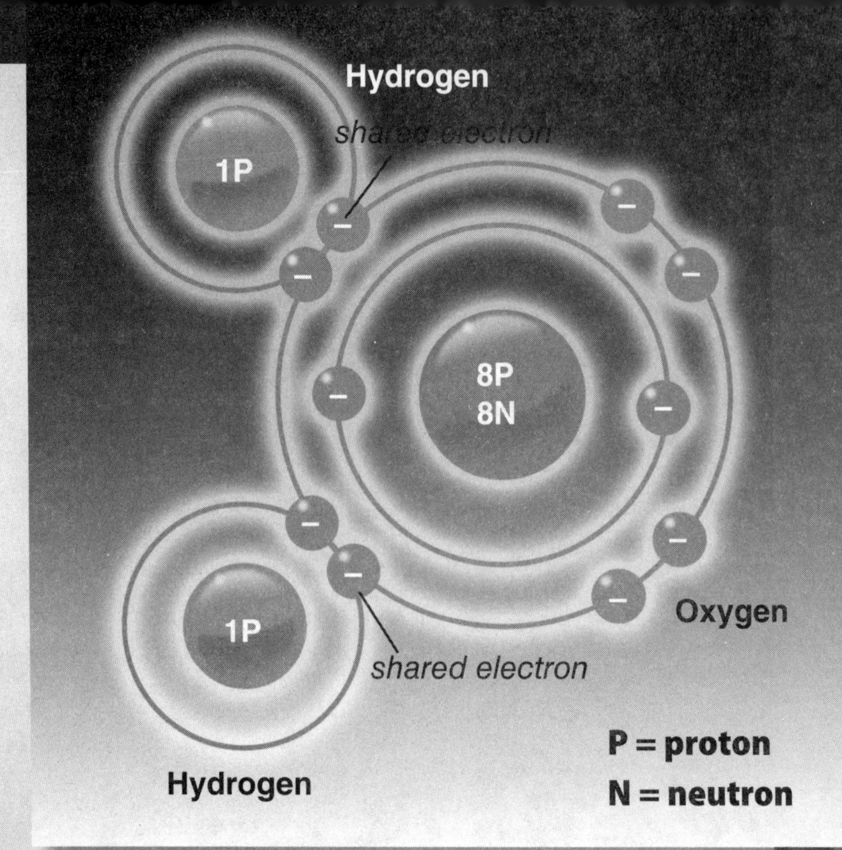

Lavoisier measured the mass of each substance before it reacted. Mass is the amount of matter a substance has. Then, he measured the mass of the new substance after the reaction. The masses were the same! He proved that matter is never lost or gained. The total mass stays the same even when something new is made. He called this law Conservation (kon-ser-VAY-shuhn) of Mass. To conserve is to keep the same. Matter can change, but its mass stays the same.

Oxidation

Lavoisier found that oxygen reacts with many things. It often reacts in the same way. Oxygen rusts iron. It tarnishes copper. It turns other metals gray. Why? Lavoisier called this oxidation (OX-id-ay-shun). Sadly, he died before he found out what it really was.

Scientists now know that oxygen atoms have a "hole." The hole is in the electrons on the outside of the atom. These holes suck in electrons from other atoms. Then the oxygen drags the other atoms. The atoms are stuck together. This makes new compounds. The rust that Lavoisier saw was one of these compounds.

Comprehension Question

How is oxidation a chemical reaction?

Chemical Reactions

Lavoisier

Antoine Laurent Lavoisier (ahn-TWAN loh-RAHN la-vwah-ZYAY) is considered the "Founder of Modern Chemistry." During his life as a scientist, Lavoisier did some important new work. The work and how he did it changed chemistry forever.

The Element Oxygen

Lavoisier was the first to identify and name oxygen as an element. He defined an element as a substance that could not break down into a simpler substance. That definition remains today. Oxygen is one important element. It is in the air we breathe, and water is made partly of oxygen, as well.

Modern scientists also know that an element is made up of the same kind of atoms. That means that each little part of the element is made of the same mix of three particles. The atoms are made of protons, neutrons, and electrons. The different mix of particles makes different kinds of atoms. Different kinds of atoms make different elements.

Lavoisier named 33 other substances as elements, too! There are over 100 different elements in the universe, and many of them were not known until recently.

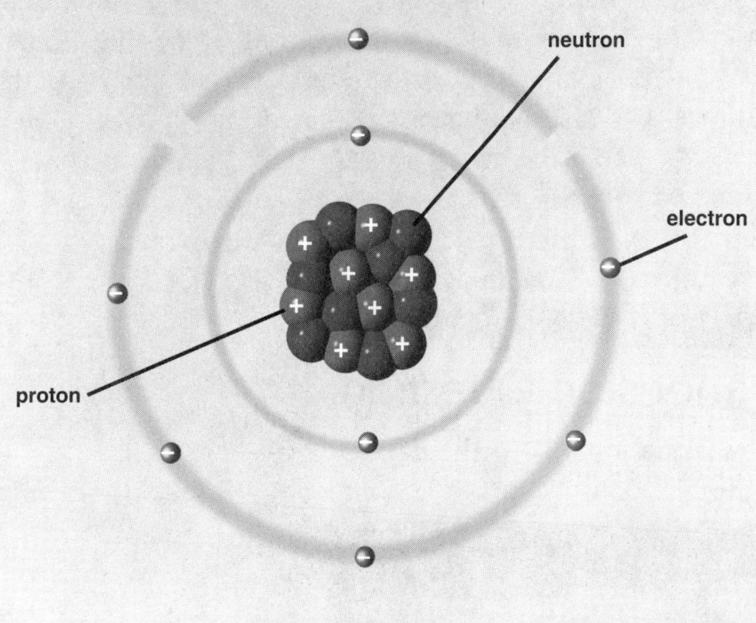

Oxygen atom

Law of Conservation of Mass

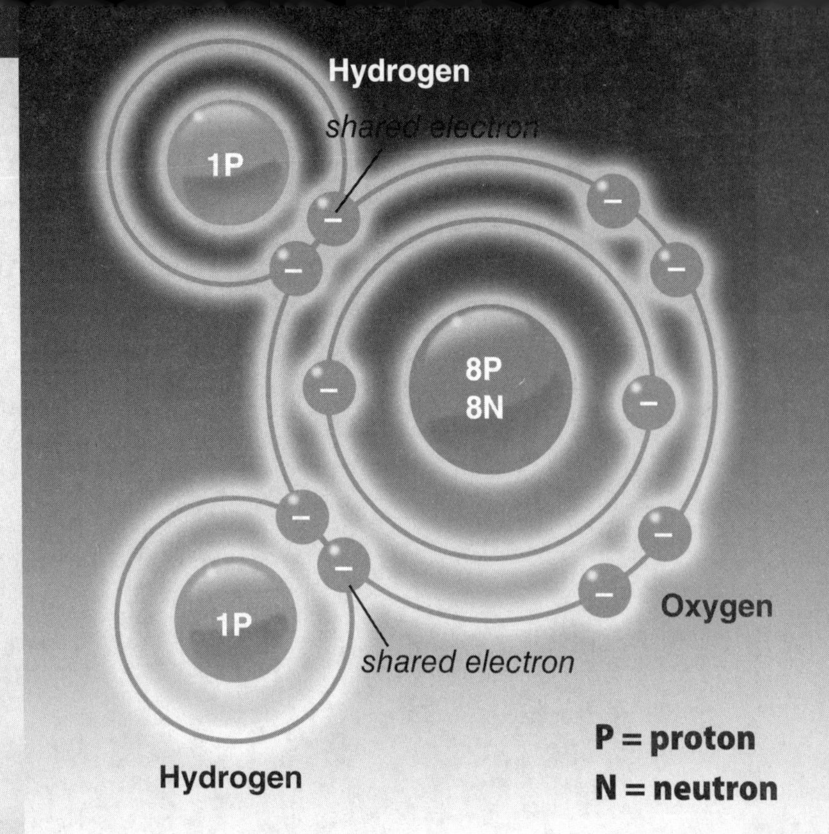

Lavoisier worked on chemical reactions. A chemical reaction happens when one or more substances react and turn into a new substance. For example, when oxygen gas and iron metal come together, there is a reaction. They become rust.

Lavoisier measured the mass of each substance before it reacted. Mass is the amount of matter a substance has. Then, he measured the mass of the new substance after the reaction. The masses were the same! He proved that matter is never lost or gained. Even when something new is made, the total mass stays the same. He called this law Conservation of Mass. To be conserved is to remain the same. Matter can change, but its mass always stays the same.

Oxidation

Lavoisier also discovered that oxygen reacts with a number of other elements in the same way. Oxygen rusts iron, tarnishes copper, and turns other metals gray. What was happening? Lavoisier called the reaction oxidation (OX-id-ay-shun). Sadly, he died in the French Revolution before he found out what it really was.

Modern scientists know that oxygen atoms have a "hole" in the electrons on the outside of the atom. These holes suck in electrons from other atoms. Then the oxygen drags the other atoms along for the ride. The atoms are stuck together. This creates new compounds. The rust that Lavoisier studied was one of these compounds.

Comprehension Question

Describe how oxidation follows the Law of Conservation of Mass.

Energy

Did you just sit down? You needed energy to do it. Did you brush your teeth this morning? Did you ride in a car, bus, or train today? Energy made that happen. In fact, all day, every day, energy is on the move. It is in you. It is around you. It is everywhere.

What is energy? Energy is being able to do work. It doesn't matter who or what is doing the work. Energy is being used. Machines use it. The sun uses it. Living things use it, too. That means you use it all the time.

Energy cannot be made. It cannot be destroyed. It is always the same amount. It may change form. It may start as motion. It may turn into heat. But the amount will not change.

Energy is found in many forms. It can be heat. It can be light. It can be motion. There are many more. All of them belong to one of two types. First there is kinetic (ki-NET-ik) energy. Then there is potential (puh-TEN-shuhl) energy.

Kinetic Energy

First there is kinetic energy. It comes from motion. A bouncing ball has it. Water in waves has it. Even atoms have it. They are always moving.

Heat comes from moving atoms. They can bump other atoms. Those atoms start moving. The heat spreads. This is called thermal (THUR-muhl) energy. Sound makes things vibrate. That is movement, too.

Electrical (ee-LEK-tri-kuhl) energy is also motion. It comes from electrons when they move. That is electricity. When this happens in a storm, we call it lightning.

Potential Energy

The second type is potential energy. This type is stored. It comes from position. Is nothing moving? The energy is stored. No work is being done. The energy is ready to be used later.

Where does potential energy come from? It can come from gravity. Think of a ball at the top of a hill. Think of all the water behind a dam. They both have energy. If you let them, the ball and the water will move. That is the energy stored in them. It comes from where they are.

Where can potential energy be? It can be in small things, too. Chemical (KEM-ih-kuhl) energy has it. It is stored in the bonds between atoms. It comes out when the atoms move. Nuclear (NOO-klee-er) energy also has it.

Comprehension Question

What are the two types of energy?

Energy

Did you just sit down? You needed energy to do it. Did you brush your teeth this morning? Did water flow from the faucet? Energy did that, too. Did you ride in a car, bus, or train today? Energy made that happen. In fact, all day, every day, energy is on the move. It is in you. It is around you. It is everywhere.

What is energy? Energy is being able to do work. No matter who or what is doing the work, energy is being used. Machines use energy. The sun uses energy. Living things use energy, too. That means you use energy all the time.

Energy cannot be created. It cannot be destroyed. It is always the same. It may change form. It may be motion and turn into heat. But the total amount of energy will not change.

Energy is found in many forms. Heat, chemical energy, light, and motion are all forms of energy. There are many more. All of them belong to one of two types. First there is kinetic (ki-NET-ik) energy. Then there is potential (puh-TEN-shuhl) energy.

Kinetic Energy

The first type is kinetic energy. It comes from motion. A bouncing ball has it. Water in waves has it. Even atoms have it. They are always moving.

Heat is energy that can spread. It comes from the moving atoms. It is called thermal (THUR-muhl) energy. Sound energy is sound waves making things vibrate.

Electrical (ee-LEK-tri-kuhl) energy is also moving. It comes from electrons. They are tiny parts of atoms. They can move from atom to atom. That is electricity. If this happens in a storm, we call it lightning.

Potential Energy

The second type is potential energy. This type is stored. It comes from position. Is nothing moving? The energy is stored. No work is being done. The energy is ready to be used later.

Potential energy can come from gravity. Think of a ball at the top of a hill. Think of all the water behind a dam. They both have energy. If you let them, the ball and the water will move. This is because the energy comes from where they are.

Potential energy can be in small things, too. Chemical (KEM-ih-kuhl) energy has it. It is stored in the bonds between atoms. It comes out when the atoms move. Nuclear (NOO-klee-er) energy also has it. It is stored in the nucleus (NOO-klee-us) of an atom. It is comes out when the nucleus is broken apart.

Comprehension Question

What is the difference between the two types of energy?

Energy

Did you just sit down? You needed energy to do it. Did you brush your teeth this morning? Did water flow from the faucet? Energy did that, too. Did you ride in a car, bus, or train today? Energy made that happen. In fact, all day, every day, energy is on the move. It is in you. It is around you. It is everywhere.

What is energy? Energy is being able to do work. No matter who or what is doing the work, energy is being used. Machines use energy. The sun uses energy. Living things use energy, too. That means you use energy all the time.

Energy cannot be created. It cannot be destroyed. It is always the same. It may change form. It may be motion and turn into heat. But the total amount of energy will not change.

Energy is found in many forms. Heat, chemical energy, light, and motion are all forms of energy. There are many more. All of them belong to one of two types. First there is kinetic (ki-NET-ik) energy. Then there is potential (puh-TEN-shuhl) energy.

Kinetic Energy

The first type is kinetic energy. It is the energy of motion. Water in waves has this energy. Atoms, molecules, and all things in motion all have it. Even you!

Heat is energy that can spread across objects. It is called thermal (THUR-muhl) energy. It comes from the atoms moving and vibrating. Sound energy is sound waves making things vibrate.

Electrical (ee-LEK-tri-kuhl) energy is also moving. Electrons are tiny parts of atoms. When they move from atom to atom, this energy is released. That is electricity. When this happens in a storm, we call it lightning.

Potential Energy

The second type is potential energy. This type is stored. It comes from position. Is nothing happening? The energy of the system is stored. No work is being done. The energy is ready to be used later.

Potential energy can come from gravity. It is caused by position. A ball at the top of a hill has this energy. Water behind a dam has it. The energy comes from where they are. If allowed to, the ball and the water will move because of energy.

Potential energy isn't just big things ready to move. Chemical (KEM-ih-kuhl) energy has it, too. It is stored in the bonds between atoms. Nuclear (NOO-klee-er) energy also has it. It is stored in the nucleus (NOO-klee-us) of an atom. It holds the nucleus together. It is released when the nucleus is smashed apart or into another nucleus.

Comprehension Question

Compare and contrast the two types of energy.

Energy

Did you just sit down? You needed energy to do it. Did water flow from the faucet when you brushed your teeth this morning? Energy did that, too. Did you ride in a car, bus, or train today? Energy made that happen. In fact, all day, every day, energy is on the move in you, around you, and everywhere else.

What is energy? Energy is the ability to do work. No matter who or what is doing the work, energy is being used. Machines use energy. The sun uses energy. Living things use energy, too. That means you use energy all the time.

Energy cannot be created or destroyed. The total amount of energy that goes into a system must equal the total energy out of that system. It may change in form within the system, but the total amount of energy will not change.

Energy is found in many forms. Heat, chemical energy, light, and motion are all forms of energy. There are many more. All the different forms of energy belong to one of two types. There are kinetic (ki-NET-ik) energy and potential (puh-TEN-shuhl) energy.

Kinetic Energy

The first group is kinetic energy. Kinetic energy is the energy of motion. Water in waves or currents has this energy. Atoms, molecules, and all things in motion all have it.

Heat is the kind of kinetic energy that can be transferred between objects. It is called thermal (THUR-muhl) energy. The movement of atoms and molecules in matter causes this energy. Sound energy, or sonic energy, is made when vibrating movement creates sound waves. Electrical (i-LEK-tri-kuhl) energy is the movement of electrical charge. Electrons are tiny particles inside atoms. When electrons move, this energy is released. It is called electricity. Lightning is an example of this kind of energy.

Potential Energy

The second group is potential energy. Potential energy is stored or caused by position. If nothing is happening and no work is being done, the energy of the system is stored. It is ready to be used in the future.

Chemical (KEM-ih-kuhl) energy is an example of potential energy. It is stored in the bonds of atoms and molecules. Nuclear (NOO-klee-er) energy is stored in the nucleus (NOO-klee-us) of an atom. It holds the nucleus together. It is released when the nucleus is split or joined with another nucleus.

Potential energy from gravity or gravitational energy is caused by position. A ball resting at the top of a hill has this energy. Water has it when it is behind a dam. If allowed to, the ball and the water will move because of energy.

Define the 2 types of energy.

Compare and contrast them,

Heat

Heat doesn't stay put. It moves. It gets passed from one thing to another. This idea may sound very simple. There are some big ideas behind it. The study of heat is called thermodynamics (thur-moh-dye-NAM-iks).

All things are made of matter. Matter is made of tiny parts. They are called atoms. No one can see an atom. That takes a very strong microscope. All things are made of atoms even if you can't see them.

Heat is all about moving atoms. Atoms are always moving. They shake back and forth. When they move more, they have more energy. The more energy they have, the more heat they have. The atoms in hot things move a lot. The atoms in cold things move less.

Heat moves in three ways. They are conduction, convection, and radiation.

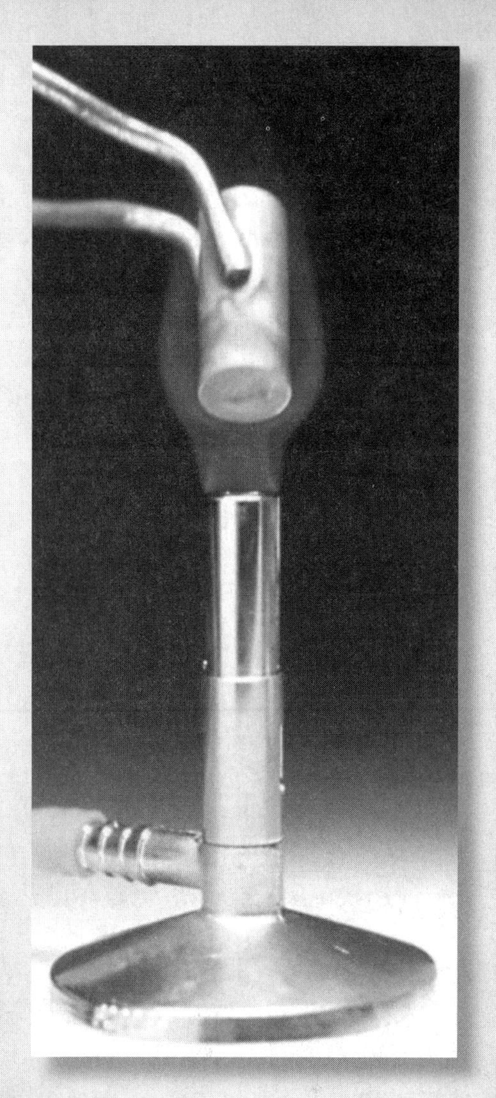

Conduction

Have you ever touched a pot on the stove? Did you burn yourself? If you have, that's conduction. Conduction is when the heat energy moves. Hot atoms pass heat to cold atoms. The atoms in the pot are moving a lot. You touch your hand to the pot. The atoms in the pot smash into the atoms in your hand. The atoms in your hand start moving more. This heats them up. Your hand gets very hot!

Convection

Have you ever been inside a cool room? Did you open a door to a hot day? If you have, that's convection. Convection is when hot matter moves. It brings heat with it. You open the door. Hot air from outside comes through the door. The air is hot. You can feel it on your cool skin. You go out and close the door. The hot air still got in the cool room. It mixes with the cool air. The air in the room gets warmer. That is because of the hot air.

Radiation

Have you ever lain out in the sun? Did you get hot? If you have, that's radiation. Radiation doesn't move through atoms. You lie out in the sun. You can feel the heat from the sun. There are no atoms in space between the sun and Earth. The sun's heat can not get to us through conduction. It can not use convection, either. It uses radiation. The atoms in the sun are full of energy. They shoot off energy called photons (FOH-tawns). The photons travel through space to Earth. They hit atoms here. Then the energy in the photon moves the atoms. This heats them up.

Now you know how heat works. Think big. Think of the whole universe. It is lots of little atoms dancing. They move back and forth. They knock into each other. They pass heat energy along. Welcome to the dance!

Comprehension Question

How does heat move?

Heat

Heat doesn't just stay in one place. It moves. It gets passed from one thing to another. This idea may sound very simple, but there are some big ideas behind it. The study of heat is called thermodynamics (thur-moh-dye-NAM-iks).

Everything in the universe is made of matter. Matter is made of tiny particles called atoms. No one can see an atom. That takes a very powerful microscope. Everything is made of atoms, even if you can't see them.

Heat is all about how atoms move. Atoms are always moving. They vibrate back and forth. The more they move, the more energy they have. The more energy they have, the hotter they are. The atoms in hot things vibrate very quickly. The atoms in cold things vibrate very slowly.

Heat moves in three ways. They are conduction, convection, and radiation.

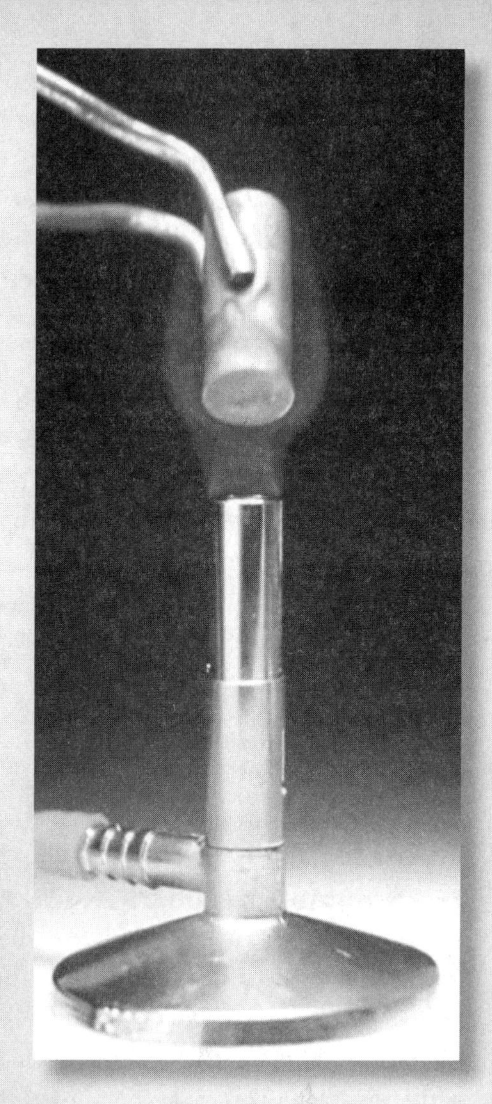

Conduction

Have you ever touched a pot on the stove and burned yourself? If you have, that's conduction. Conduction is when the heat energy of some atoms passes to other atoms. The atoms in the pot are vibrating very fast. You touch your hand to the pot. The atoms in the pot smash into the atoms in your hand. The atoms in your hand start vibrating faster. This heats them up. Suddenly your hand is very hot!

Convection

Have you ever been inside a cool room and then opened the door to a hot day? If you have, that's convection. Convection is when hot material moves into a cooler area and brings heat with it. When you open the door, a wave of hot air goes into the room. You can feel its heat on your skin. You might go outside and close the door. That hot air is still inside the cool room. It mixes with the colder air. The air in the room gets a little bit warmer because of the new, hot air.

Radiation

Have you ever lain out in the sun on a hot day? If you have, that's radiation. Radiation doesn't move through atoms. Lying out in the sun, you can feel the heat from the sun. There are no atoms in space between the sun and Earth. The sun's heat cannot get to us through conduction. It cannot use convection, either. It gets to the Earth through radiation. The atoms in the sun are so full of energy that they shoot off packets of energy called photons (FOH-tawns). The photons travel through space to Earth. They hit atoms here. Then the energy in the photon moves the atoms, heating them up.

Now you know how heat works. You can think of the whole universe as lots of little atoms dancing back and forth. They knock into each other and pass heat energy along. Welcome to the dance!

Comprehension Question

Describe the three ways heat moves.

Heat

The study of heat and how it moves is called thermodynamics (thur-moh-dye-NAM-iks). Heat doesn't just stay in one place. It moves. It gets passed from one thing to another. This may sound very simple, but there are some big ideas behind this simple concept.

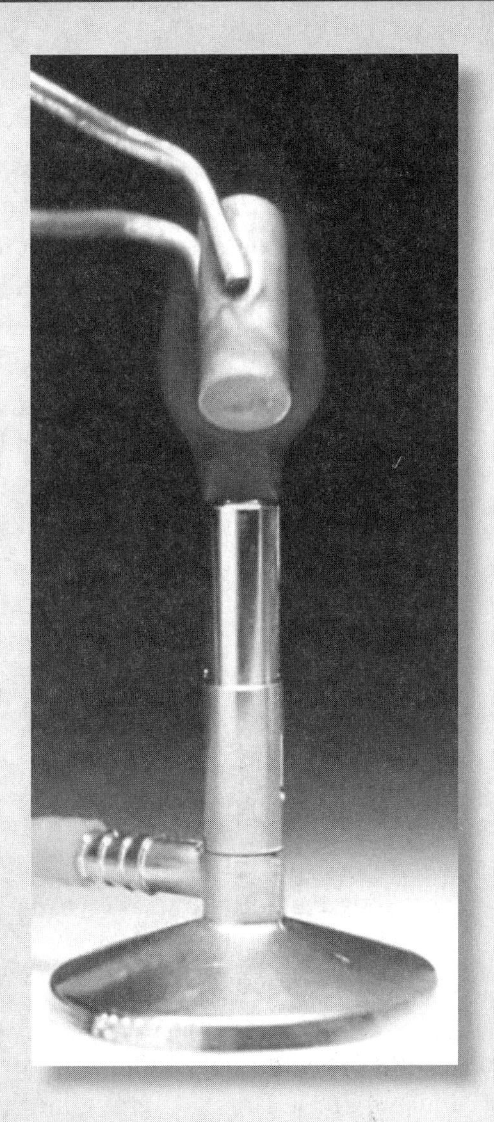

Matter is the "stuff" that makes up the contents of the universe. Matter is made of tiny particles called atoms. No one can see an atom. That takes a very powerful microscope.

But atoms are the basis of matter, no matter what.

How the atoms move has a lot to do with heat. Atoms are always moving, vibrating back and forth. The more they move, the more energy they have. The more energy they have, the hotter they are. The atoms in hot things are vibrating very quickly. The atoms in cold things are vibrating very slowly. Heat moves in three ways. They are conduction, convection, and radiation.

Conduction

Have you ever tried to pick up a pot from the stove and burned yourself? If you have, you've experienced conduction. Conduction is when the heat energy of some atoms passes over to other atoms. The atoms in a pot on the stove are vibrating very fast. If you pick up the pot, the atoms in the pot smash into the atoms in your hand. The atoms in your hand start vibrating faster. This heats them up. Suddenly your hand is very hot!

Convection

Have you ever been inside a cool room and then opened the door to a hot day? If you have, you've experienced convection. Convection is when hot material moves into a cooler area and brings heat with it. At the door, a wave of hot air would have swept into the room. You would have felt its heat on your skin. Even after you went outside and closed the door, that hot air is still inside the cool room. It mixes with the colder air. The air in the room gets just a little bit warmer because of the new, hot air.

Radiation

Have you ever lain out in the sun on a hot day? If you have, you've experienced radiation. Radiation is when energy moves in waves instead of vibrating atoms. Lying out in the sun, you can feel the heat from the sun. There are no atoms between the sun and Earth. The sun's heat cannot get to us through conduction or through convection. The sun's heat still gets to Earth, though, and it does that through radiation. The atoms in the sun are so full of energy that they shoot off packets of energy called photons (FOH-tawns). The photons travel through space until they hit atoms on Earth. Then the energy in the photon sets the atoms moving, heating them up.

Now that you know how heat works, you can think of the whole universe as lots of little atoms, all dancing back and forth, knocking into each other, and passing heat energy along. Welcome to the dance!

Comprehension Question

Compare and contrast the three ways heat moves.

Heat

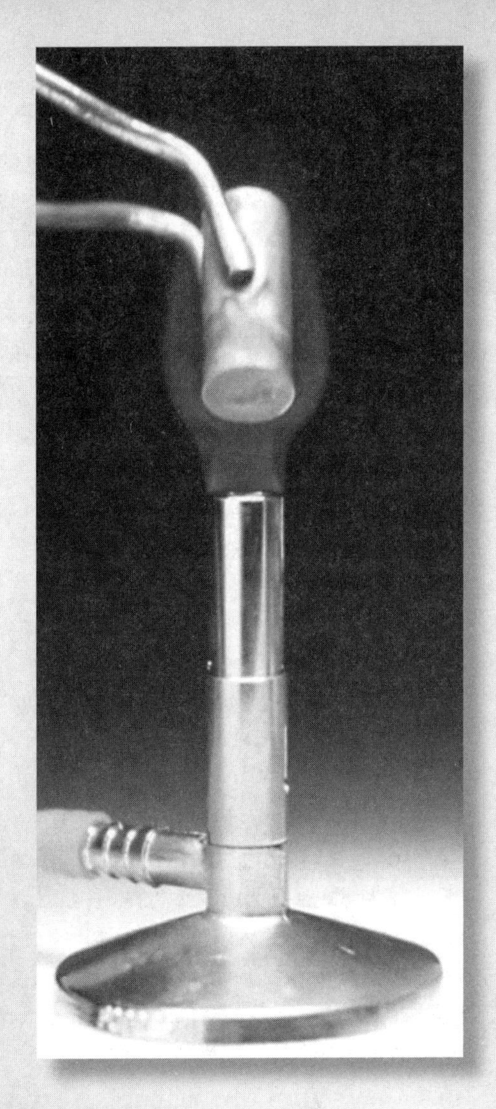

The study of heat and how it moves is thermodynamics (thur-moh-dye-NAM-iks). Heat is always moving. In fact, heat is motion: atoms vibrating with heat energy. That energy can pass from one object to another. This may sound like a very simple concept, but there are some big ideas behind it.

Matter is the "stuff" out of which the universe is built. Most matter is composed of tiny particles called atoms. It takes a very powerful electron microscope to see an atom. Whether or not you can see them, atoms are the basis of matter, no matter what.

Heat is the movement of atoms; they are perpetually vibrating back and forth. That movement is energy, and the more they move, the more energy they have. The more energy they have, the hotter they are. Scientists say that the atoms are excited. Hot objects are composed of excited atoms vibrating very quickly; cold objects are composed of atoms vibrating very slowly.

Heat can also move from object to object, in three ways. They are conduction, convection, and radiation.

Conduction

If you have ever burned yourself on a hot pot, you've experienced conduction. Conduction is the passage of heat energy from hot, excited atoms to cooler, slower atoms. The atoms that compose the pot are vibrating very fast. When your hand makes contact with the pot, the atoms in your hand contact the atoms of the pot. The excited atoms in the pot collide with the atoms in your hand. The atoms in your hand are excited and start vibrating faster. Suddenly your hand is very hot!

Convection

If you have ever opened a door from a cool room to a hot day outside, you've experienced convection. Convection is the passage of hot material into a cooler area. The hot material is composed of excited atoms, and brings that heat energy with them. When you open the door, a wave of hot air and excited atoms sweep into the room. Even after you went outside and closed the door, that hot air is still inside the cool room, mixing with the colder air. The air in the room gets just a little bit warmer because of the new, hot air.

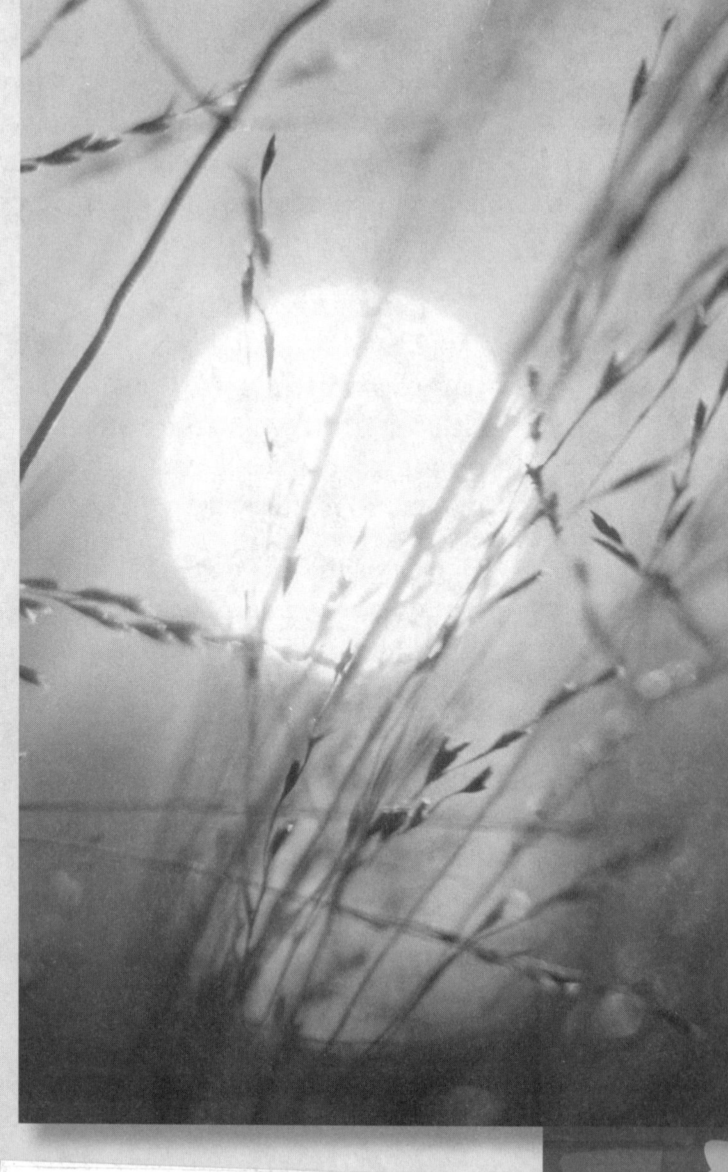

Radiation

If you have ever lain out in the sun on a hot day, you've experienced radiation. Radiation is when energy moves in waves instead of vibrating atoms. Lying out in the sun, you can feel the heat from the sun. With no atoms between the sun and the Earth, that heat cannot get to us through conduction or through convection. However, the sun's heat still gets to the Earth, and it does that through radiation. The atoms in the sun are so excited that they shoot off packets of energy called photons (FOH-tawns). The photons travel through space until they hit atoms on Earth. The energy in the photon sets the atoms moving, heating them up.

Now that you know how heat works, you can think of the whole universe as lots of little atoms, all dancing back and forth, knocking into each other, and passing heat energy along. Welcome to the dance!

Comprehension Question

How are the three means of heat transfer similar and different?

Explain the 3 ways heat moves.

Define thermo-dynamics

Use matter and atoms in a sentence.

Sunlight

A World of Energy

Think of a nice day at the park. You park your car. You spread out your blanket. Little white clouds drift across the sky. There are trees growing up out of the ground. There are birds flying from tree to tree. It may be peaceful. But this park is full of energy!

What makes the clouds move? How did the clouds get into the sky to start? The trees are doing lots of hard work, too. They are growing taller and making seeds. The birds use energy all day long. They fly. They eat. They fly away from cats. Even your car used energy to get here.

Where does all the energy come from? Here's a hint. Look into the sky!

Energy from the Sun

There is a lot of energy at the park. Almost all of it came from the sun. The sun is a giant ball. It is made of very hot gas. That energy comes to us through heat and light. That is called radiation (RAY-dee-ay-shun). The sun has lots of energy. It shoots out extra energy. It shoots out in little packets called photons (FOH-tawns). The photons travel through space. Then they hit Earth.

Some of the photons hit the air in the sky. The air heats up. The air on the side of the planet closer to the sun heats up more than the other side. Hot air gets bigger. Cold air gets smaller. So the hot air spreads out. It spreads into where the cold air is. This makes wind.

Some of the photons hit water in the ocean. The water heats up. Some of it heats up a lot. It evaporates. It turns into water vapor. The water vapor rises up into the sky. It hits cold air high up in the sky. It condenses. The water becomes clouds!

Some of the photons hit chlorophyll (KLOR-oh-fill) in the tree leaves. The photons have energy. The leaves grab it. They use the energy. They make food for the tree. The tree uses the food to grow taller. It uses it to make new seeds and fruit.

Some of the photons hit the birds. The birds don't use those photons much. Instead, they eat the seeds and fruit from the trees. The seeds and fruits are filled with energy. That energy came from photons from the sun. The birds use the energy in their food to fly, to grow, and to make chicks.

Your car used energy to get to the park. That energy came from the sun, too. It happened in the time of the dinosaurs. The plants used photons from the sun to store energy back then, too. Then the plants died. They got buried. But the energy was still in the plants. The plants got squeezed and squeezed. It took millions of years. Over time, the plants turned into oil. Your car burns the oil to drive.

Comprehension Question

Where does most of the energy on Earth come from?

Sunlight

A World of Energy

Imagine a nice day at the park. You park your car and find a place to spread out your blanket. Little wispy clouds drift across the sky. There are trees growing up out of the ground and birds flying from tree to tree. It may be peaceful, but that park is full of energy!

What makes the clouds move across the sky? How did the clouds get up there in the first place? The trees are doing lots of hard work growing taller and making seeds. The birds use energy all day long flying, feeding, and staying away from cats. Even the car you drove in used energy to get here.

Where does all the energy come from? Hint: look into the sky!

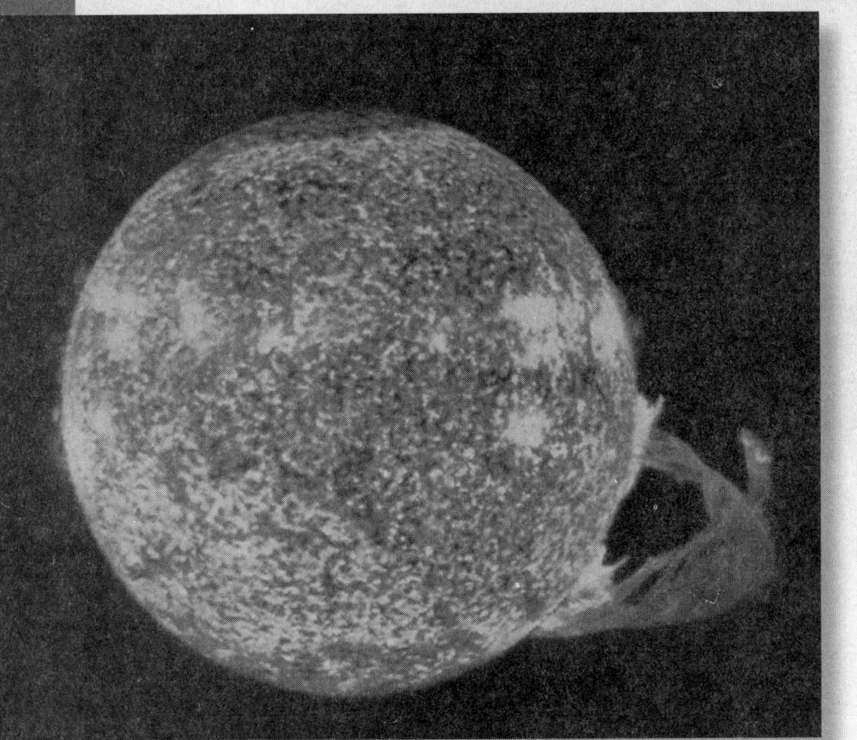

Energy from the Sun

Believe it or not, almost all the energy at the park came from the sun. The sun is a giant ball of very hot gases. It has a lot of energy! That energy comes to us through heat and light radiation. The sun has so much energy, it shoots out little packets of energy called photons (FOH-tawns). The photons travel through space until they fall on Earth.

Some of the photons hit air molecules in the atmosphere. Those air molecules heat up. The air on the side of the planet closer to the sun heats up more than the other side. Hot air expands. Cold air gets smaller. So the hot air spreads out to where the cold air is shrinking up. Winds are created.

Some of the photons hit water molecules in the ocean. Those water molecules heat up. Some of them heat up so much they evaporate into the air. Since they are warm, they rise up into the sky. They rise up to the colder air at the top of the atmosphere. They condense into water vapor. They become clouds!

Some of the photons hit chlorophyll (KLOR-oh-fill) molecules in the tree leaves. Those molecules capture the energy in the photons. They use the energy to make food for the tree. The tree uses that food to grow taller. It uses it to make new seeds and fruit.

Some of the photons hit the birds. The birds don't use those photons for much. Instead, they eat the seeds and fruit from the trees. The seeds and fruits are filled with energy. That energy came from photons from the sun. The birds use the energy in their food to fly, to grow, and to make chicks.

Even the car that you drove to the park in uses energy. It came from the sun, too, but a long time ago. When dinosaurs walked the Earth, the plants used photons from the sun to store energy in food. Then the plants died and got buried underground. The energy was still in the plants. The plants got squeezed and compressed over millions of years. Eventually, the plants turned into oil. The oil was turned into gasoline and pumped into your car.

Comprehension Question

Describe three ways energy from the sun is used.

Sunlight

An Energetic World

Imagine a beautiful day at the park. You park your car, find a place to spread out your blanket, and watch wispy clouds drift slowly across the sky. Trees grow and stretch their branches all around, dropping seeds to the ground. Birds twitter and fly from tree to tree. It feels very peaceful in the park, but it is full of energy!

What makes the clouds move across the sky? How did they form there in the first place? You know that the trees are working hard to grow taller and make seeds. The birds are using energy to fly, to feed, and to stay safe from predators. Even the car you drove to the park used energy to get there.

Where does all the energy come from? Hint: look up!

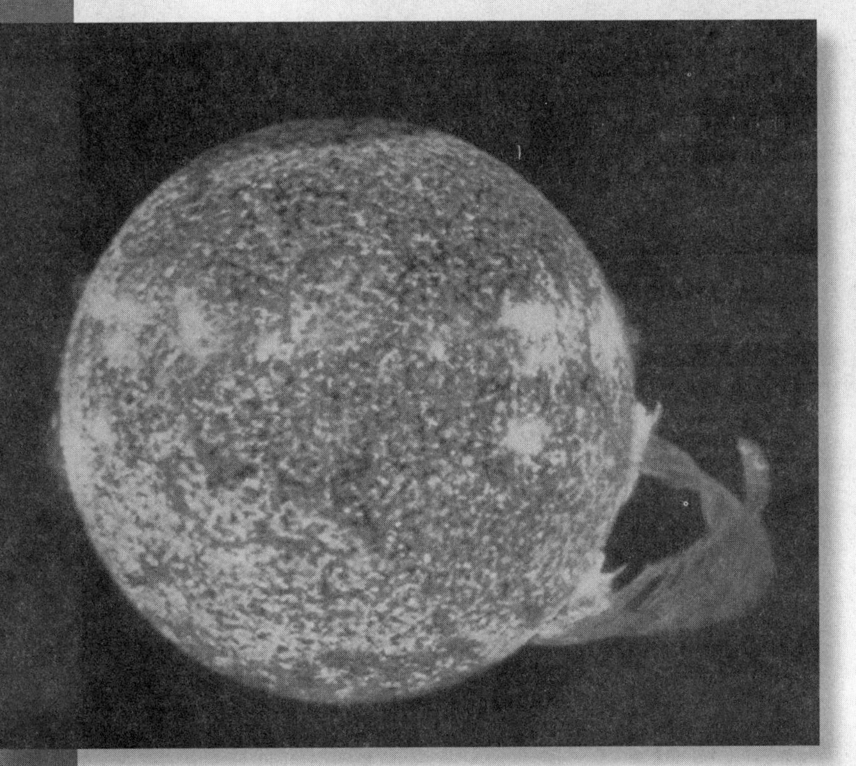

Energy from the Sun

Believe it or not, almost all the energy being used at the park came from the sun. The sun is a giant ball of incredibly hot gases with a great deal of energy. That energy comes to us through heat and light radiation. The sun has so much energy, it constantly shoots out little packets of energy called photons (FOH-tawns). The photons travel through space until they fall on planet Earth.

Some of the photons hit air molecules in the atmosphere, and then those air molecules heat up. The air on the side of the planet closer to the sun heats up more than the other side. Hot air expands, and cold air gets smaller. So the hot air spreads out to where the cold air is shrinking up. When this happens, winds are created.

Some of the photons hit water molecules in the ocean, and those water molecules heat up. Some of them heat up so much they evaporate into the air. Since they are warm, they rise up into the atmosphere. Eventually, they get up to the colder air at the top of the atmosphere where they condense into water vapor and become clouds.

Some of the photons hit chlorophyll (KLOR-oh-fill) molecules in the tree leaves. Those molecules capture the energy in the photons. They use the energy to make food for the tree. The tree uses that food to grow taller and to make new seeds and fruit.

Some of the photons hit the birds; however, the birds don't use those photons for much. Instead, they eat the seeds and fruit from the trees. The seeds and fruits are filled with energy from the sun's photons. The birds use the energy from their food to fly, to grow, and to reproduce.

Even the car that you drove to the park uses energy. That energy also came from the sun, but it came long ago. When dinosaurs walked the Earth, the plants used photons from the sun to store energy in food. When the plants died and became buried underground, the energy remained in the plants. The plants were squeezed and compressed over millions of years. Eventually, the plants turned into oil, and the oil was made into gasoline and pumped into your car.

Comprehension Question

How do we see the effects of the sun's energy every day?

Sunlight

An Energetic World

Imagine that you are enjoying a beautiful day at the park. You jump out of the car, spread your blanket under the trees, and look up at the wispy clouds drifting slowly across the sky. Trees grow everywhere around you, stretching their limbs to the sun while birds twitter and fly between the branches. It seems very calm and peaceful in the park, but it is really alive with energy!

Do you know what makes the clouds move across the sky, and how they formed there in the first place? Something makes the trees grow taller and gives them what they need to breed new seeds and fruit. The birds use something all day long to help them fly, feed, and stay safe from predators. That something is energy. Even the car you drove to the park needed energy to get there.

Where does all the energy come from? Hint: look up!

Energy from the Sun

Believe it or not, almost all the energy used at the park originated from the sun. The sun is a giant ball of incredibly hot gases with tremendous energy. That energy is transported to Earth through heat and light radiation. The sun engenders so much energy that it constantly shoots out photons, which are miniscule packets of energy. The photons travel rapidly through space until they arrive at planet Earth.

Some of the photons hit air molecules in the atmosphere, and then those air molecules become warmer. The air on the side of the planet facing the sun naturally heats up more than the side facing away. Hot air expands and cold air restricts, so the hot air spreads out to where the cold air is shrinking. When this happens, winds are created.

Some of the photons hit water molecules in the oceans and other water bodies, and those molecules become warmer. Some of them heat so much they become gaseous and evaporate. Since they are warm, they rise into the atmosphere, eventually arriving at the top where the air is colder. There they condense into water vapor and become clouds.

Some of the photons hit chlorophyll molecules that are stored in tree leaves. Those molecules capture the energy in the photons and use it to nourish the tree by producing food. The tree uses that food to grow and to generate new seeds and fruit.

Some of the photons hit the birds; however, the birds don't use those photons much. Instead, they eat the energy-filled seeds and fruit from the trees, which were energized by the photons. The birds use the energy from their food to fly, to grow, and to reproduce.

Even the car that you drove to the park uses energy to operate. That energy also came from the sun, but it came eons ago. When dinosaurs walked the Earth, the plants used photons from the sun to store energy. When the plants died and became buried underground, the energy remained in the plants, which were squeezed and compressed over millions of years. Eventually, the plants turned into oil, and the oil was converted into gasoline and pumped into your car.

Comprehension Question

Describe how the winds, the weather, plants, animals, and even machines use the energy from the sun.

Electrical Circuits

electrons move through the atoms of a wire

What are generators? How do they work? Generators give us electricity. They start with fuel. They turn the fuel into electricity. That makes light bulbs glow. It makes TVs show the news. Let's take a look at how it works.

Electricity does not move like we do. It only moves through conductors (KON-duck-tors). Conductors are used for wires. They have loose electrons. The electrons move back and forth through the wire. Copper wire is a good choice. Aluminum is, too. Even gold and silver work.

Electricity also needs a push to make it go. That is what a generator is for.

Generators use magnets. They move the magnets along a wire. The magnets push the electrons inside the wire. They do not have to touch them. This makes a flow of electrons. That is called electricity.

Think of a pipe. It is filled with Ping-Pong balls. What would happen if you pushed one more ball into the end? All the balls would move along the pipe and bump the last one out. This is like electrons moving along the wire.

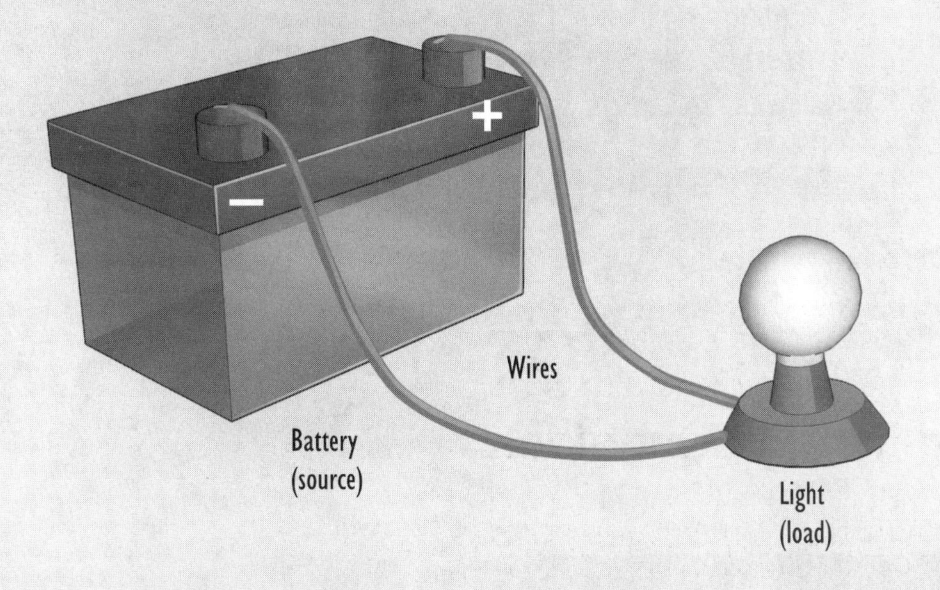

Amps and Volts

The generator's magnet pushes. The electrons go down the wire. We can measure this two ways. The magnet pushes a number of electrons. That is measured in amperes (AM-peers). It is called amps for short. The magnet also puts a force on the electrons. This is measured in volts.

Amps and volts are put together to see how much power is used. This is called watts. Watts are amps times volts. That tells you two things. It tells you how many electrons are moving. It also tells you how much force is behind them.

AC vs. DC

Thomas Edison made the first light bulb. He wanted to sell electricity. He wanted to bring it into homes. He made a promise to light up New York. There was one problem. The electricity for homes was direct current. That is called DC for short. The electrons always went the same way down the wire. DC cannot go very far. The electrons fall out of the wire. DC is not good for homes.

Nikola Tesla worked for Edison. He had an idea. He wanted to move the electrons a new way. He made electrons go back and forth very fast. They changed direction many times in one second. He called it alternating current. It is called AC for short. This let electricity go much farther.

The two men both wanted their way. Edison wanted DC. Tesla wanted AC. In the end, Tesla's ideas worked best. Today, we use AC in our homes.

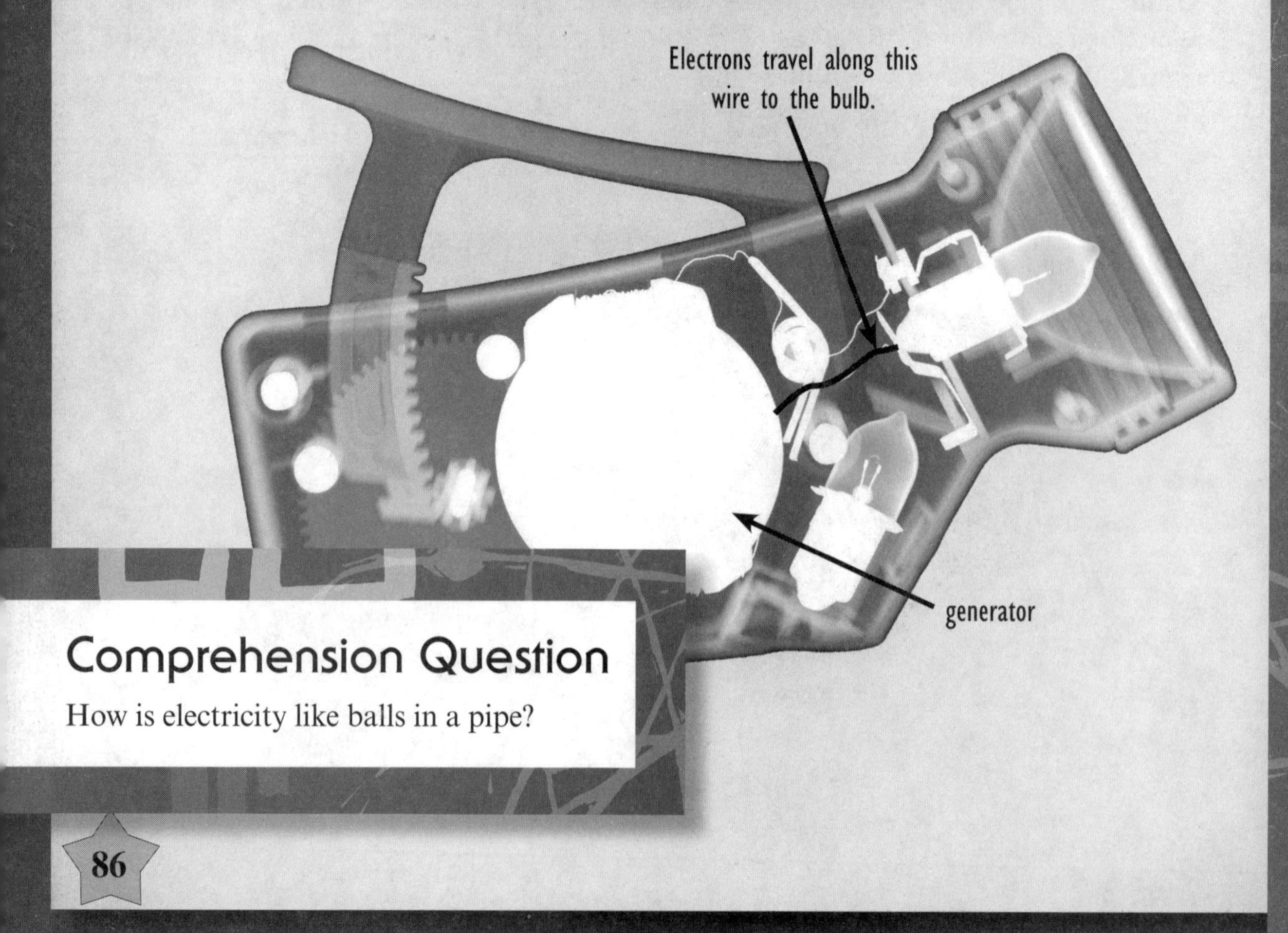

Comprehension Question

How is electricity like balls in a pipe?

Electrical Circuits

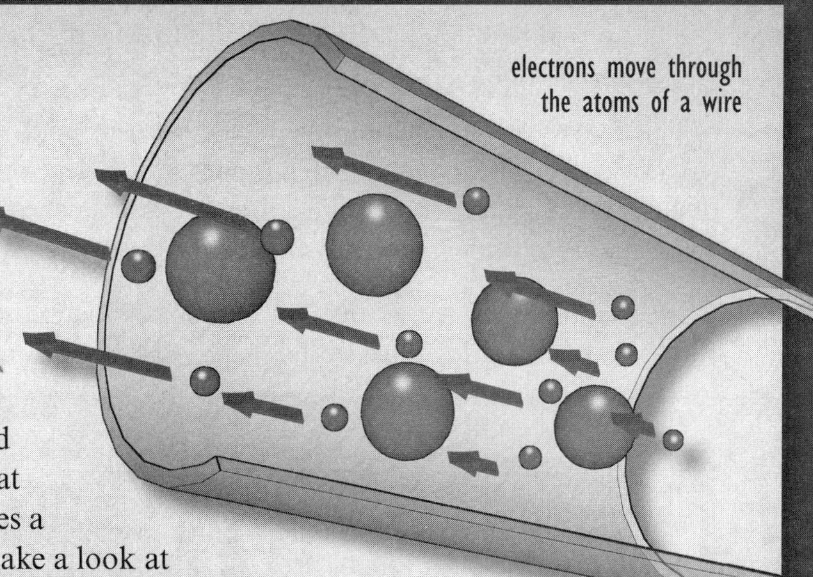

What is a generator? How does it work? Generators give us electricity. They start with fuel from other things. This is turned into electricity. Electricity is what makes a light bulb glow. It makes a television show the news. Let's take a look at how it works.

First, electricity does not move like we do. It only moves through conductors (KON-duck-tors). Conductors are used for wires. They have loose electrons. The electrons move back and forth through the wire. Copper wire is a good choice. Aluminum is, too. Even gold and silver can be used.

Electricity also needs a push to make it go. That is what a generator is for.

Generators use magnets. They move the magnets along a wire. The magnets push the electrons inside the wire. They do not have to touch them. This creates a flow of electrons. That is electricity.

Imagine a pipe. It is filled with Ping-Pong balls. What would happen if you pushed one more ball into the end? All the balls would move along the pipe and bump the last one out. This is like electrons moving along the wire.

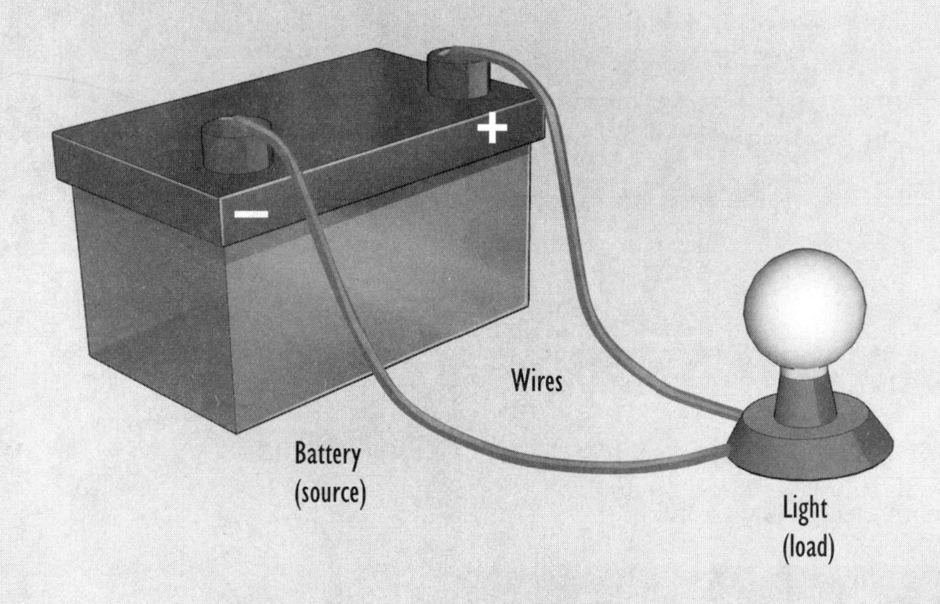

Amps and Volts

The generator's magnet pushes. The electrons go down the wire. We can measure this two ways. The magnet pushes a number of electrons. That is measured in amperes (AM-peers). It is called amps for short. The magnet also puts pressure on the electrons. This is measured in volts.

Amps and volts are put together to see how much power is used. This is measured in watts. Watts are amps times volts. Watts tell you how many electrons are moving and how much force is behind them.

AC vs. DC

Thomas Edison invented the light bulb. He wanted to sell electricity. He wanted to bring it into people's homes. He also made a promise. He said he would light up New York. There was a problem, though. The electricity for homes was direct current. That is called DC for short. The electrons always went in the same way down the wire. DC cannot travel long distances. The electrons fall out of the wire. DC is not good for homes.

Nikola Tesla worked for Thomas Edison. He had an idea. He wanted to use another kind of current. Alternating current makes electrons go back and forth very fast. They change direction many times in one second. Alternating current is called AC for short. This makes it easy for electricity to travel over long distances.

The two men both thought they had the best idea. Edison wanted DC. Tesla wanted AC. In the end, Tesla's ideas worked best. Today, we use AC in our homes.

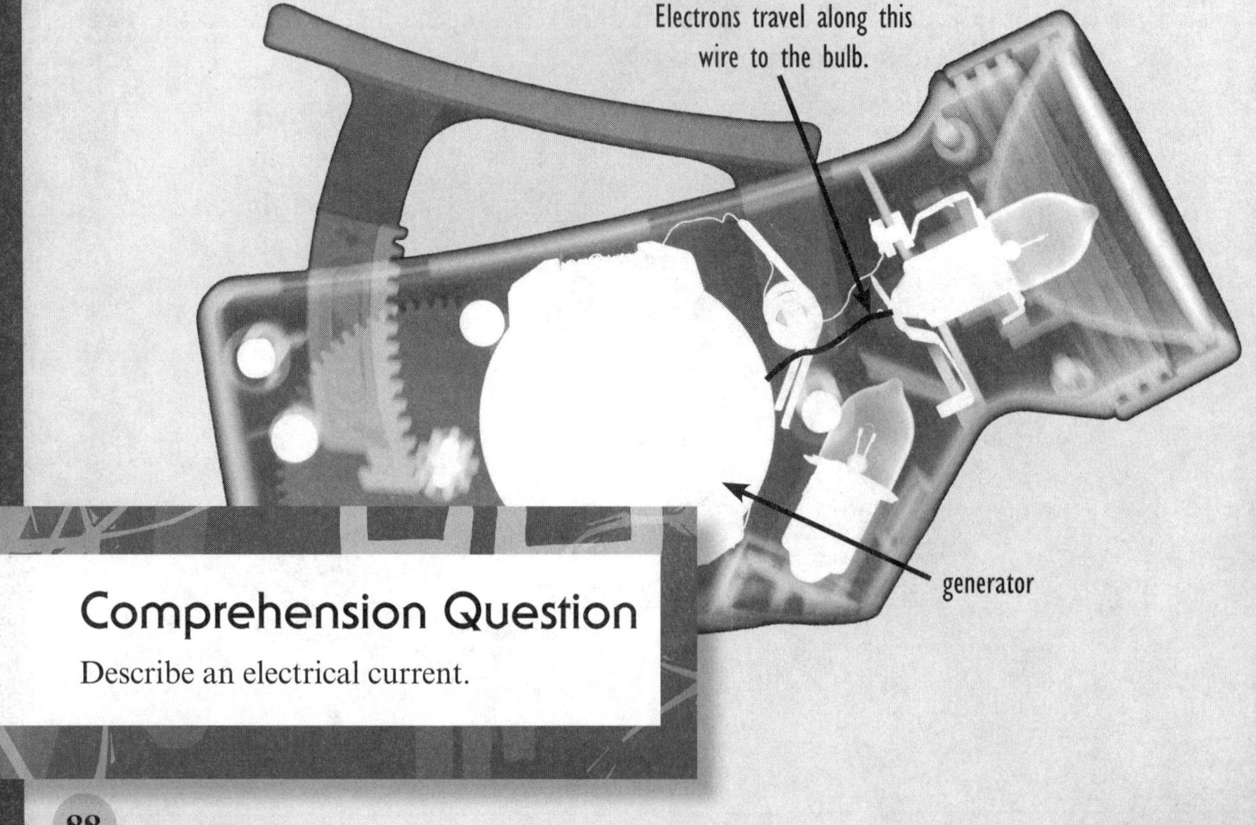

Comprehension Question

Describe an electrical current.

Electrical Circuits

electrons move through the atoms of a wire

What is a generator? How does it work? Generators give us electrical energy. It starts with energy from other things. It turns that into electricity. This energy is what makes a light bulb glow. It makes a television show the news. Let's take a look at how it works.

First, electricity moves through conductors (KON-duck-tors). Conductors are materials that share electrons easily. The electrons move between the atoms. Copper wire is a good choice. Aluminum, gold, and silver are other metals that can also be used.

Electricity also needs a push to get it moving. That is what a generator is for.

Generators move the magnets along a wire. The magnets push the electrons inside the wire. They do not have to touch them. This creates a flow of electrons. That is electricity.

Imagine a pipe. This is like the wire. It is filled with Ping-Pong balls. What would happen if you pushed one more ball into the end? All the balls would move along the pipe and bump the last one out. This is like electrons moving along the wire.

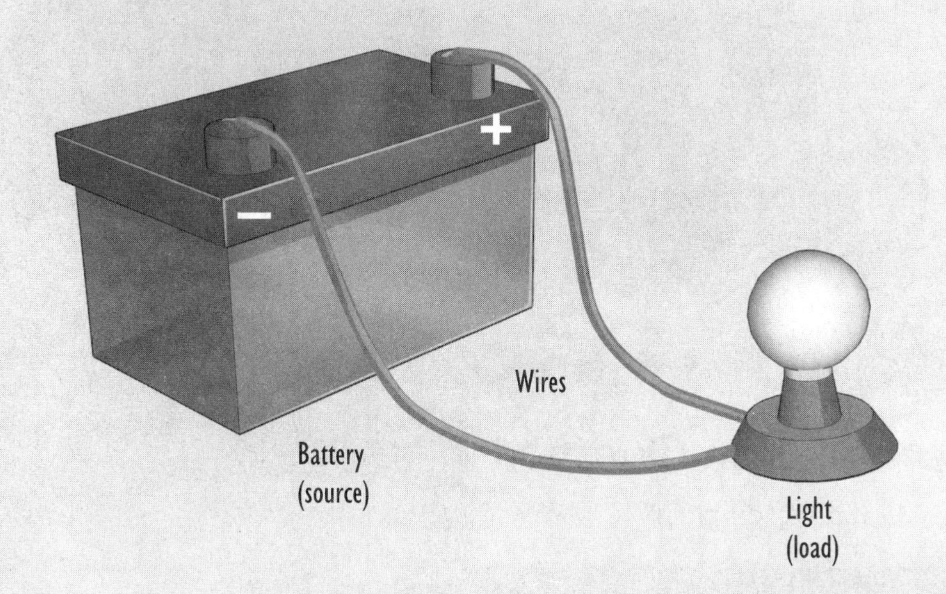

Amps and Volts

The generator's magnet pushes the electrons down the wire. Two things happen. The magnet pushes a certain number of electrons along the wire. This is measured in amperes (AM-peers). They are also called amps for short. The magnet also puts pressure on the electrons. This is measured in volts.

Amps and volts are put together to see how much power is being used. This is measured in watts. Watts are found by multiplying the amps and volts together. That tells you how many electrons are moving and how much force is behind all those electrons.

AC vs. DC

Thomas Edison invented the light bulb in 1879. He wanted to sell electricity. He wanted to bring it into people's homes. He promised to light up New York. There was a problem, though. The electricity for homes was direct current, or DC for short. The electrons always went in the same direction down the wire. DC cannot travel long distances. The electrons fall out of the wire. DC is not good for homes.

Nikola Tesla worked for Thomas Edison. He had an idea. He wanted to use another kind of current. Alternating current, or AC, makes electrons go back and forth very fast. They change direction many times in one second. This makes it easy for electricity to travel over long distances.

The two men argued. Edison wanted DC. Tesla wanted AC. In the end, Tesla's ideas worked best. Today, we use AC in our homes.

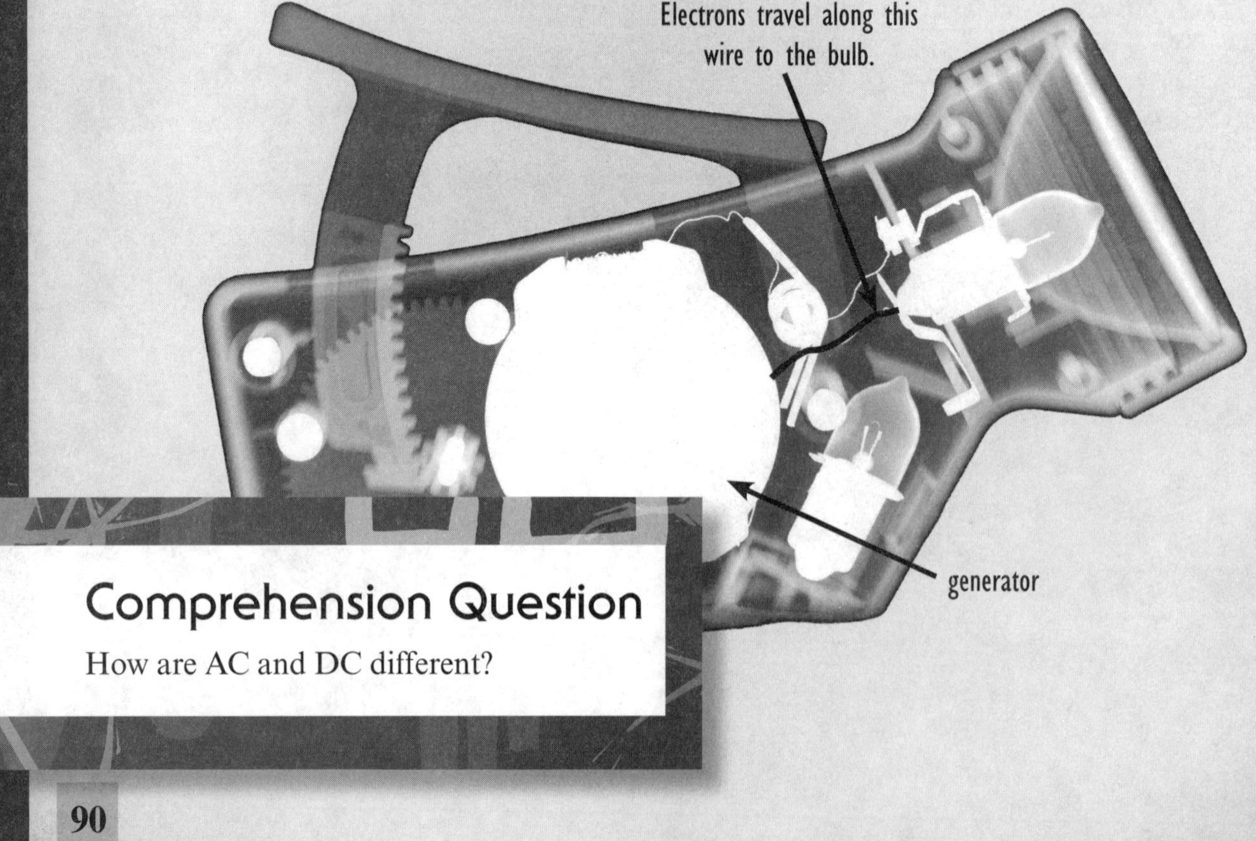

Comprehension Question

How are AC and DC different?

Electrical Circuits

electrons move through the atoms of a wire

What exactly is a generator, and how does it work? An electrical generator converts energy from other sources into electrical energy. This energy is what makes a light or a television work when switched on. To better understand how these forces work, let's take a look at how an electrical current and electromagnetism are made.

First, electricity needs a conductor to move it from one point to another. Materials that are good conductors have electrons that move easily. Copper wire is a good conductor. Aluminum, gold, and silver are other metals that are sometimes used as conductors.

Electricity also needs something to get it moving through the conductor. Generators are often used to do this.

Generators use magnets to push the electricity along a wire. The magnets push the electrons inside the wire without having to touch them. This creates a steady flow of electricity.

Imagine a pipe. This represents the wire. It is filled with Ping-Pong balls. What would happen if you pushed one more ball into the end? All the balls would move along the pipe and bump the last one out. This is like electrons moving along the wire and into an electrical appliance.

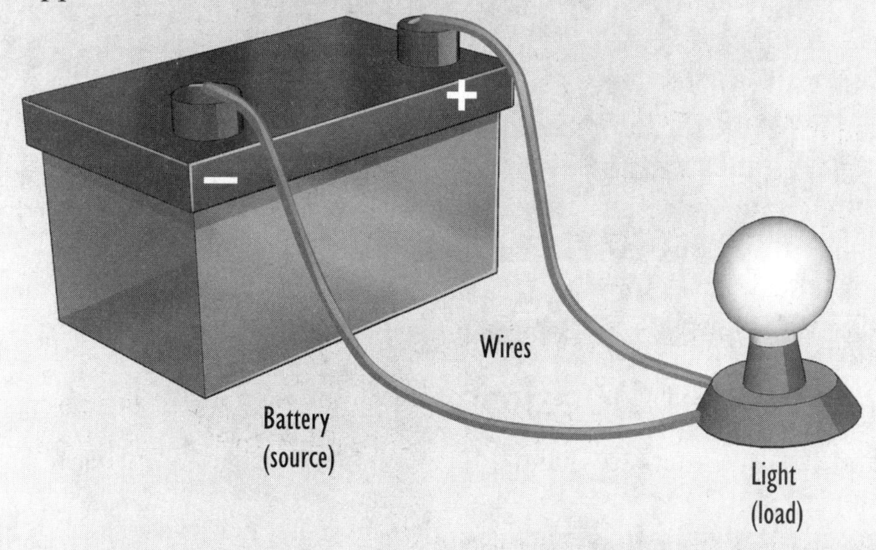

Amps and Volts

As a generator's magnet pushes the electricity along the wire, two things happen. The magnet pushes a specific number of electrons along the wire. This electrical current is measured in amperes (AM-peers), or amps. The magnet is also putting pressure on the electrons. This pressure is measured in volts.

Amps and volts are put together to determine how much power is being used. This is measured in watts. Watts are found by multiplying the amps and volts together. That tells you how many electrons are moving and how much force is behind all those electrons.

AC vs DC

After Thomas Edison invented the light bulb in 1879, he worked to bring electricity into people's homes. He also made a promise. He wanted to light up the city of New York. There was a problem, though. The electricity for homes was direct current, or DC. This is a current pushed through a circuit and flowing continuously in the same direction. Low-voltage DC currents cannot travel long distances because there are such high losses in the cables that carry them. That means DC is not good for powering our homes.

Nikola Tesla, who worked for Thomas Edison, had an idea. He wanted to use another kind of electric current. Alternating current, or AC, changes direction back and forth many times in one second. This makes it easy for electricity to travel over long distances.

The two men didn't agree. In the end, Tesla's ideas worked best. Today, the electricity that comes from a power plant and is used in our homes is AC.

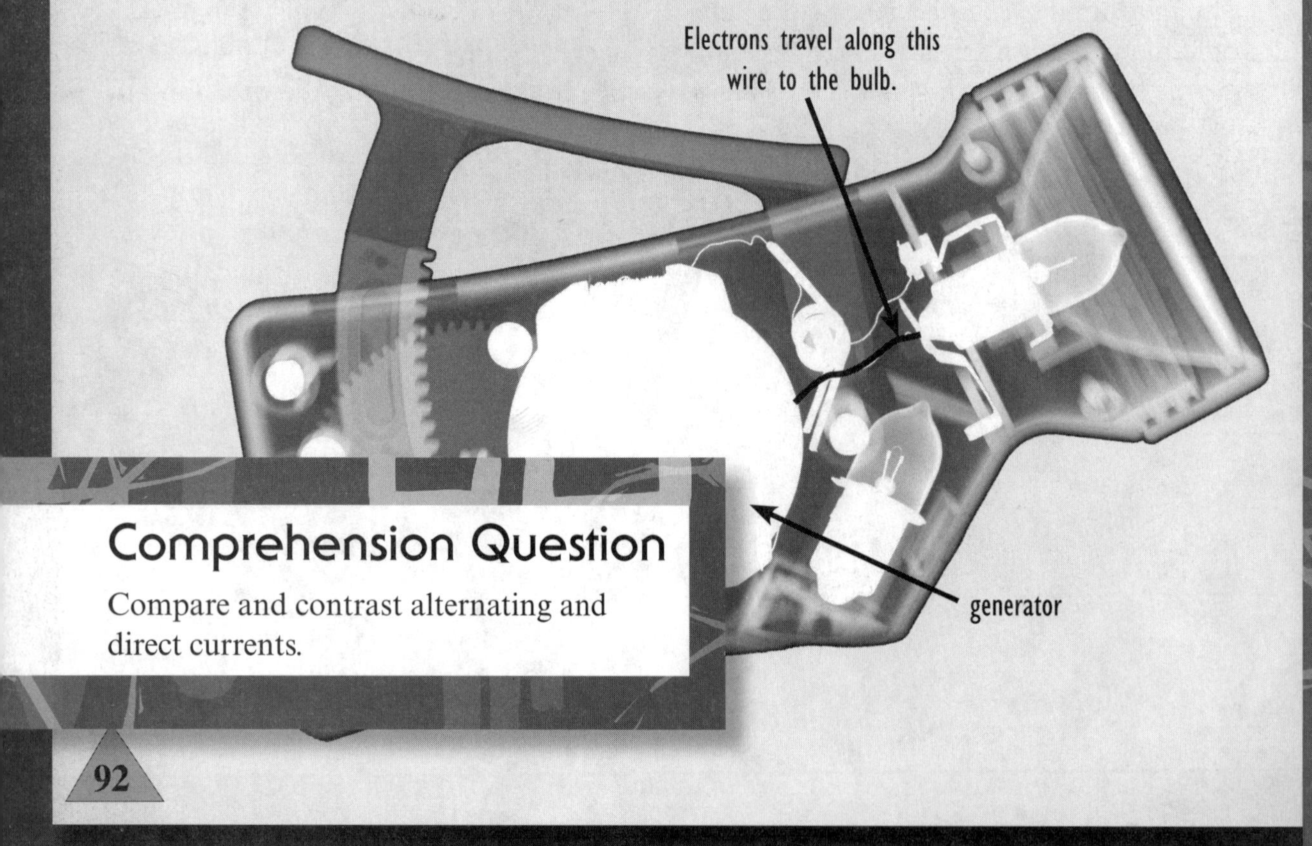

Comprehension Question

Compare and contrast alternating and direct currents.

Vibrations

Put your fingers on your throat. Hum. What do you feel? Your can feel vibrations (vie-BRAY-shuns) in your throat. When you use your voice your body makes noise. Your voice box vibrates.

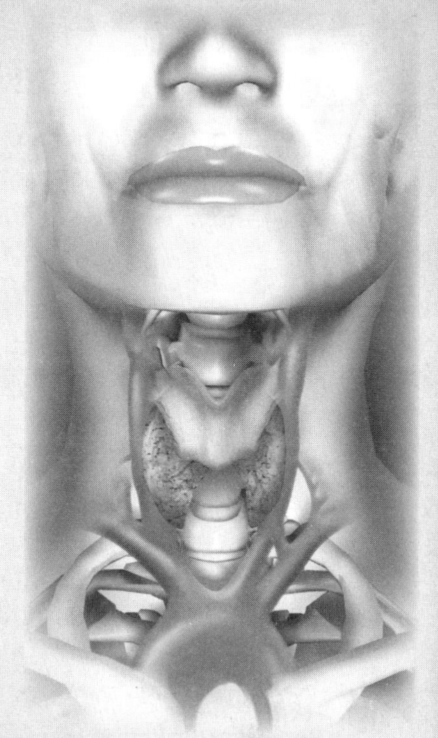
the voice box

Sound comes from vibrations. The atoms in the air are always moving. They can move in a pattern. That creates waves. The waves move through matter. They cause vibrations. The vibrations are picked up by our ears. Our brain turns them into sounds that we hear.

Sound Waves

Sounds waves are compression (KOM-presh-un) waves. They are made of moving atoms. Those atoms get pushed by other atoms.

Think of a big movie blast. It starts small. It pushes outward. It pushes the air around it farther out. That air pushes more air. A sound wave is just like that. It is a vibration of atoms. They push other atoms. The other atoms vibrate the same way as the first atoms.

Not all sound waves are alike. One sound wave makes one sound. Another sound wave makes another sound. Sound waves differ in three ways:

- Wavelength is the length from wave to wave.

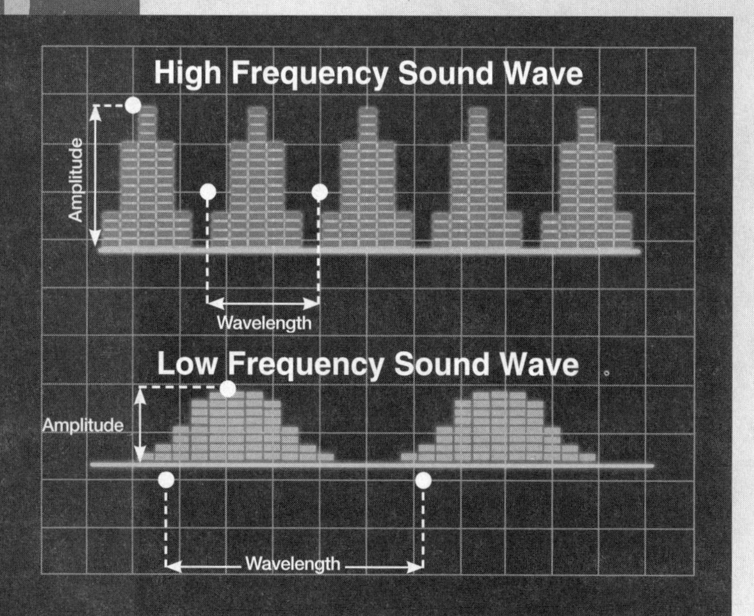

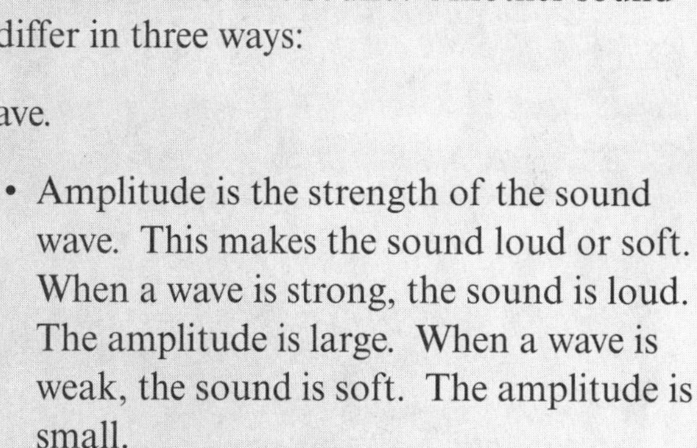

- Amplitude is the strength of the sound wave. This makes the sound loud or soft. When a wave is strong, the sound is loud. The amplitude is large. When a wave is weak, the sound is soft. The amplitude is small.

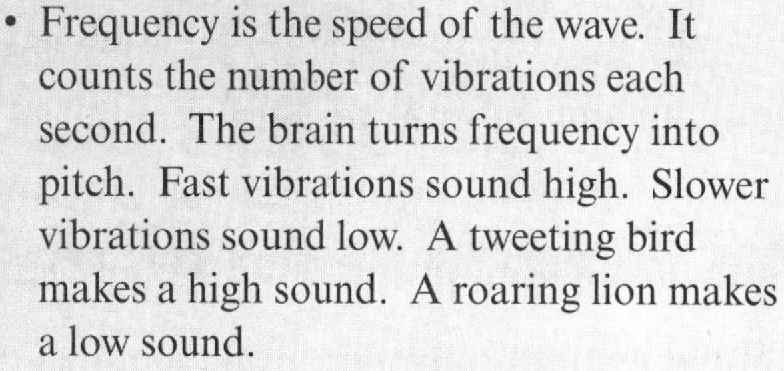

- Frequency is the speed of the wave. It counts the number of vibrations each second. The brain turns frequency into pitch. Fast vibrations sound high. Slower vibrations sound low. A tweeting bird makes a high sound. A roaring lion makes a low sound.

The Speed of Sound

Sound waves pass through everything. They go through gas like the air. They go through liquid like the ocean. They even go through solids like the ground. The atoms vibrate along with the sound wave. The speed of sound changes when it goes through different things. Sound waves move:

- slowly through gases.
- more quickly through liquids.
- fastest through solids.

Heat also affects the speed of moving sound waves. Hot things let sound move faster.

Sound travels about 343 meters (1,125 feet) per second. That is like traveling 1,217 kilometers (756 miles) per hour!

Bats use sound to sense the world around them.

Comprehension Question

What is sound?

Vibrations

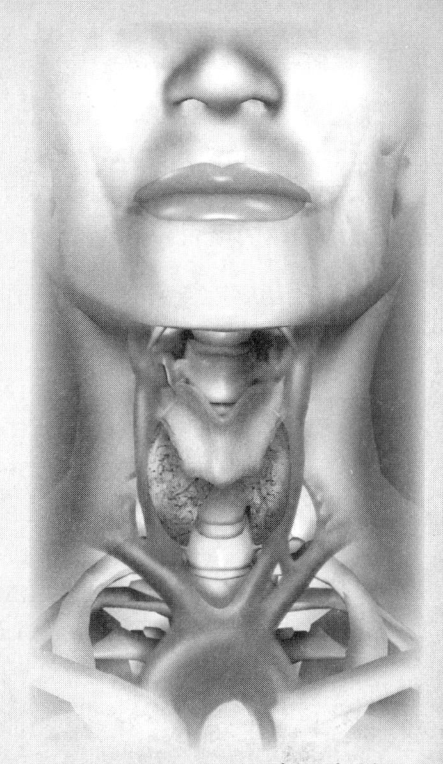
the voice box

Put your fingers on your throat and hum. What do you feel? Your can feel vibrations in your throat when you use your voice. This is because your body makes noise–your voice–by vibrating your voice box.

Sound comes from vibrations. The atoms and molecules that make up the air are always moving. When they move in a pattern, it creates waves. As the waves move through matter, they cause vibrations. The vibrations are picked up by the ear and sent as messages to the brain. The brain turns them into the sounds we hear.

Sound Waves

Sounds waves are compression (KOM-presh-un) waves. That means that they are made of atoms being pushed, or compressed, by other atoms. Think of a big movie explosion. It starts small, but it pushes outward. That pushes the air around it farther out. That air pushes more air. A sound wave is just like that. It is a vibration of atoms pushing more atoms to vibrate the same way.

Not all sound waves are alike. The differences let us hear various sounds. Scientists have discovered that sounds and sound waves differ in the following ways:

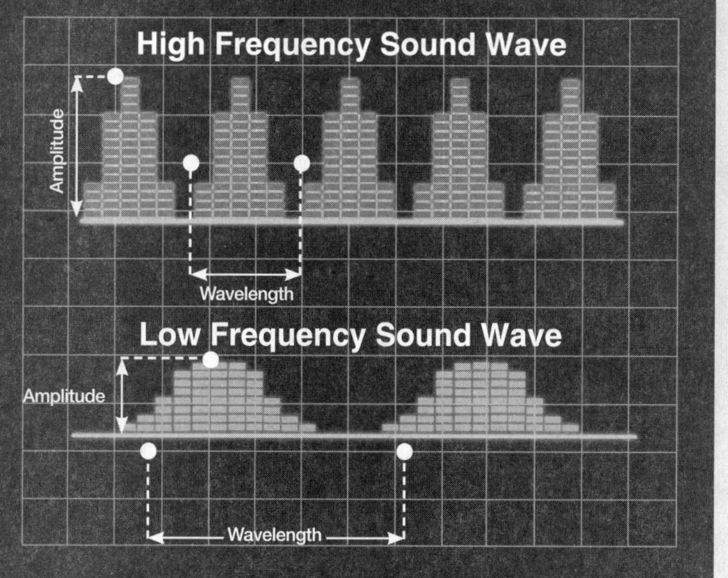

- Wavelength is the length from wave to wave.
- Amplitude is measured in the height of the sound wave. It relates to the loudness or softness of a sound. When a wave is high, the sound is loud. The amplitude is large. When a wave is low, the sound is soft. The amplitude is small.
- Frequency of sound relates to speed. The number of vibrations per second is the frequency. The brain turns frequency into pitch. Fast vibrations cause high pitch. Slower vibrations make lower-pitched sounds. A tweeting bird makes a high-pitched sound. A roaring lion makes a low-pitched sound.

The Speed of Sound

Sound waves pass through everything. They go through gases like the air. They go through liquid like the ocean. They even go through solids like the ground. The atoms vibrate along with the sound wave. The speed of sound changes when it goes through different things. Sound waves move:

- slowly through gases.
- more quickly through liquids.
- fastest through solids.

Heat also affects the speed of moving sound waves. Hot things let sound move faster.

Sound travels about 343 meters (1,125 feet) per second. That is like traveling 1,217 kilometers (756 miles) per hour!

Bats use sound to sense the world around them.

Comprehension Question

How are vibrations and sound related?

Vibrations

Put your fingers on your throat and hum. What do you feel? Your can feel vibrations in your throat when you use your voice. This is because your body makes noise–your voice–by vibrating membranes in your voice box.

Sound comes from vibrations. The atoms and molecules that make up the air are always moving. When they move in a pattern, it creates waves. As the waves move through matter, they cause vibrations. The vibrations are picked up by the ear and sent as impulses to the brain. The brain translates them as the sounds we hear.

the voice box

Sound Waves

Sounds waves are compression (KOM-presh-un) waves. That means that they are made of atoms being pushed, or compressed, by other atoms. Think of a big movie explosion. It starts small, but it pushes outward. That pushes the air around it farther out. That air pushes more air. A sound wave is just like that. It is a vibration of atoms pushing more atoms to vibrate the same way.

Not all sound waves are alike. The differences let us hear various sounds. Scientists have discovered that sounds and sound waves differ in the following ways:

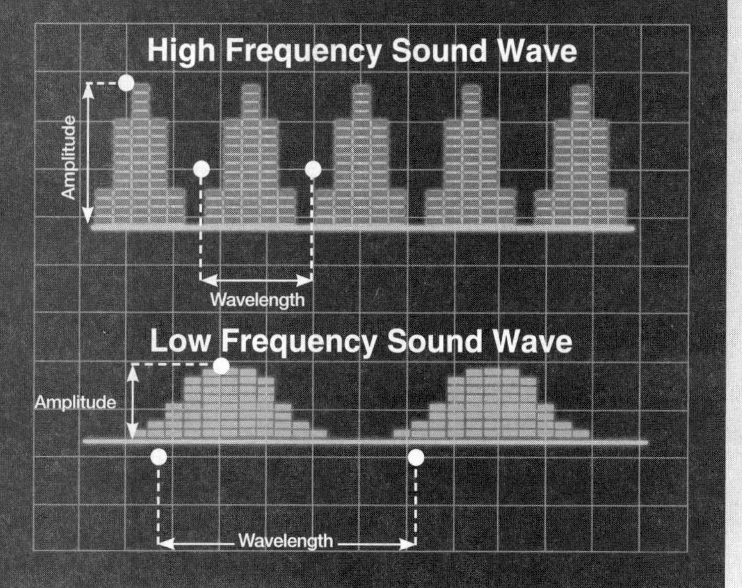

- Wavelength is the distance between the troughs on either side of a single wave.
- Amplitude is measured in the height of the sound wave. It relates to loudness or softness of a sound. When a wave is high, the sound is loud and the amplitude is large. When a wave is low, the amplitude is small and the sound is soft.
- Frequency of sound relates to speed. The number of vibrations per second is the frequency. The brain understands frequency as pitch. Fast vibrations cause high pitch. Slower vibrations make lower-pitched sounds. A tweeting bird makes a high-pitched sound. A roaring lion makes a low-pitched sound.

The Speed of Sound

Sound waves pass through all forms of matter. These include gas, liquid, and solid. The atoms vibrate because of the sound wave's vibrations. The speed of sound changes as the waves pass these different states of matter. Sound waves move:

- slowly through gases.
- more quickly through liquids.
- fastest through solids.

Temperature also affects the speed of moving sound waves. Higher temperatures cause sound to move faster. At normal room temperatures, sound travels about 343 meters (1,125 feet) per second. That is like traveling 1,217 kilometers (756 miles) per hour!

Bats use sound to sense the world around them.

Comprehension Question

What are a sound wave's wavelength, amplitude, and frequency?

Vibrations

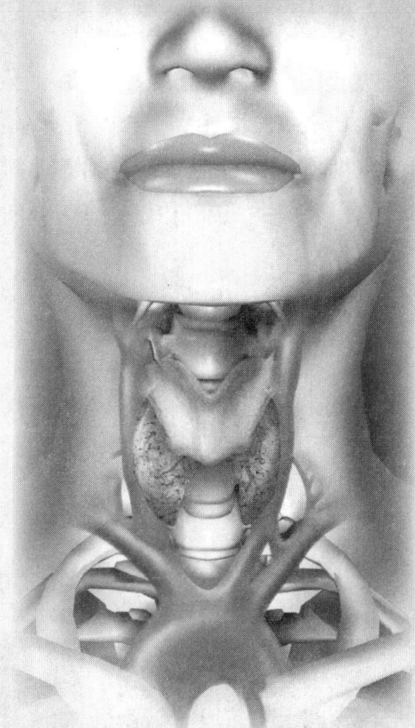
the voice box

When you put your fingers on your throat and hum, what do you feel? Your throat vibrates whenever you use your voice. In order to make noise, your body vibrates membranes in a special organ called the voice box.

Sound is vibrations. The atoms and molecules of the air are always moving, and when they move in a pattern, they create waves. As the waves move through matter, they cause vibrations. The vibrations are picked up by the ear and transmitted as impulses to the brain, which translates them into the sounds we hear.

Sound Waves

Sounds waves are compression (KOM-presh-un) waves: they are made of waves of atoms being compressed by other atoms. Like a big movie explosion, it may start small, but it pushes outward. That pushes the air around it farther out, and that air pushes more air. A sound wave is a vibration of atoms pushing more atoms to vibrate the same way.

Different sound waves create different sounds. Scientists have discovered that sounds and sound waves differ in the following ways:

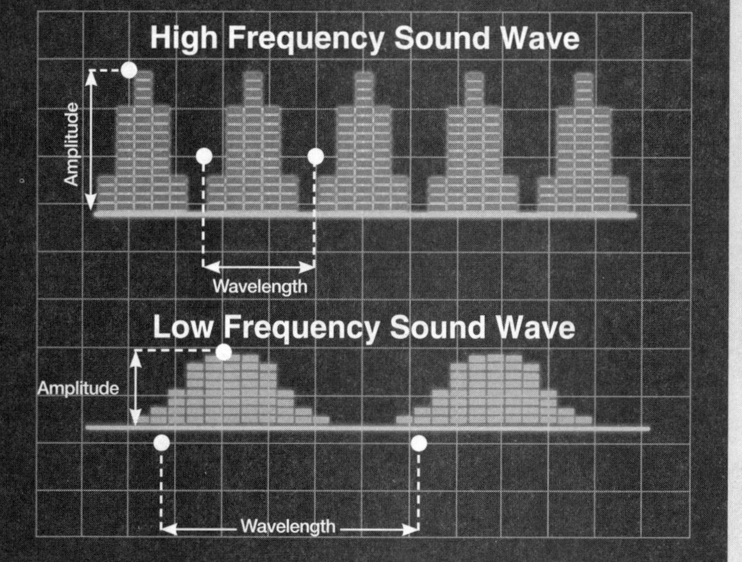

- Wavelength is the distance between the each wave of vibrating atoms.
- Amplitude is the height of the sound wave. It determines the loudness or softness of the sound. When a wave is high, the sound is loud and the amplitude is large. When a wave is low, the amplitude is small and the sound is soft.
- Frequency of sound is the speed of the wave: the number of vibrations per second. The brain understands frequency as pitch. Fast vibrations cause high pitch. Slower vibrations make lower-pitched sounds. A tweeting bird makes a high-pitched sound. A roaring lion makes a low-pitched sound.

The Speed of Sound

The vibrations of sound waves penetrate all states of matter, including gas, liquid, and solid. No matter what state they are in, the atoms still vibrate because of the sound wave's vibrations. However, the speed of sound changes as it penetrates different states of matter. Sound waves move:

- slowly through gases.
- more quickly through liquids.
- fastest through solids.

Temperature also affects the speed of moving sound waves. Higher temperatures cause sound to move faster. At normal room temperatures, sound travels about 343 meters (1,125 feet) per second. That is like traveling 1,217 kilometers (756 miles) per hour!

Bats use sound to sense the world around them.

Comprehension Question

Describe how sound waves are related to vibrations.

Radiant Light

Light is a kind of energy. It is called radiant energy. That means light sends out rays or waves. Our eyes can only see some radiant energy. Visible light is light that we can see.

Light lets us see things and people. Light waves bounce off things and the waves travel to our eyes. Our eyes work with our brain. Our eyes turn the light into what we see.

Light travels in waves. It is like water moving in waves. Waves can be different from each other in length, in rate, and in size. There is a word for each. Wavelength is length. Frequency (FREE-kwen-see) is rate. Amplitude (AM-pli-tood) is size. The color of the light comes from the rate of the wave.

Refract, Reflect, Absorb

Light shoots through the world. Then it hits things. Then something happens.

- The light can change direction. It refracts.
- The light can bounce off. It reflects.
- The light can get stuck. It is absorbed.

Light rays can hit transparent (trans-PAIR-uhnt) things. That means you can see through them. The light bends. This is called refraction (ruh-FRAK-shuhn).

Light rays can hit other things. Some of the light bounces off. It goes in a new direction. This is called reflection (ruh-FLEK-shuhn). Smooth things, like glass, reflect a lot of light. Things that are bumpy reflect less.

Light can hit a very smooth thing. Then all of the rays reflect the same. A mirror is very smooth. You can see yourself in it. Most of the light bounces right back at you. Light can hit less smooth things. Then the rays bounce off in many directions. This is why it is impossible to see your reflection in paper. Look close. The surface is bumpy. The light does not bounce right back.

Color

Color is all about absorption (uhb-ZORP-shuhn). Look at the clothes you have on. What colors are they? You may think the color is in the clothing. That is not how it works. The colors come from light. The light reflects off the clothes. Then it comes to our eyes. We see the colors of the light.

You know that light is made of waves. Each color has its own frequency. White light has the light of all the colors. It has all the frequencies in it. White light hits things. Each frequency does its own thing. Some frequencies get absorbed. They are not seen. Some reflect. They travel to our eye. We only see light that reflects.

White light is made of all the colors. Those colors are in a rainbow, too. The colors are red, orange, yellow, green, blue, indigo, and violet. Some people know the colors as ROYGBIV. Now, look at something red. The object absorbs the frequencies for OYGBIV. Only the R, or red, is reflected. The red light gets to your eye. You see the object as red.

The big idea is that the color is not in the object. It is in the reflected light.

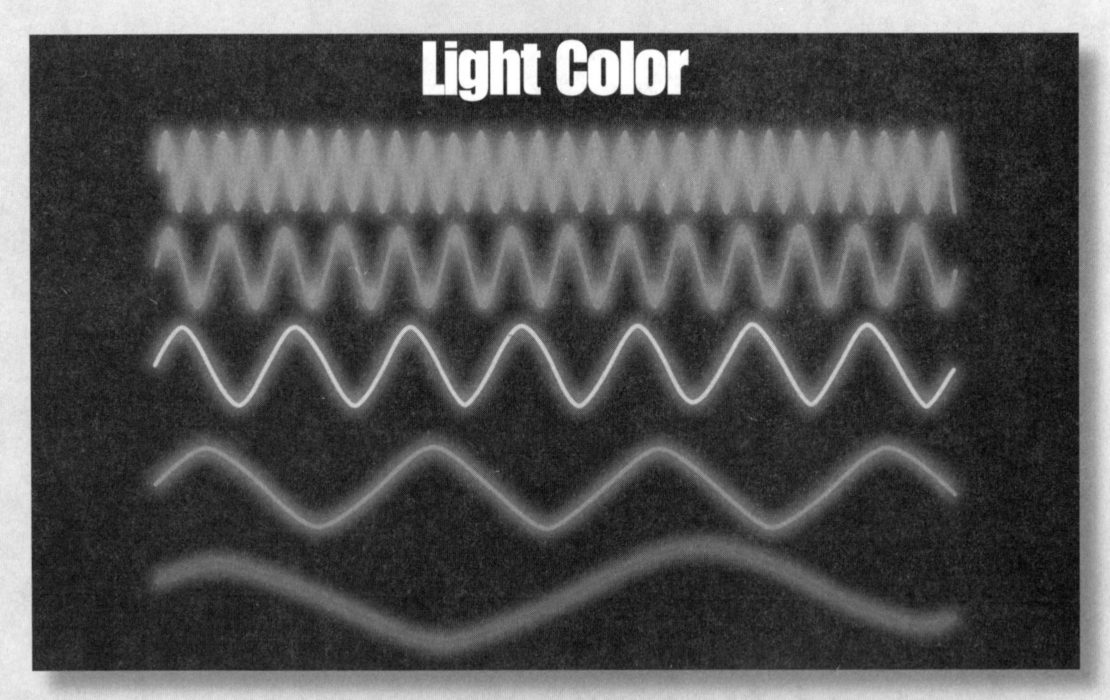

frequencies and colors of light

Comprehension Question

What three things does light do when it hits things?

Radiant Light

Light is a kind of energy. It is called radiant energy. To radiate means to send out rays. We can only see some radiant energy. We call this visible light. Visible means that we can see it.

We can see because of light. Light waves bounce off objects. Then they travel to our eyes. Our eyes and brain work together. They turn that light into what we see.

Light travels in waves. It is like water moving in waves. The color of the light comes from the energy in the wave. Waves differ from each other. They differ in length, rate, and size. There is a special word for each of these. Wavelength is length. Frequency (FREE-kwuhn-see) is rate. Amplitude (AM-pli-tood) is size. Frequency makes the color of the light.

Refract, Reflect

What happens when a light wave hits something? Several things might happen.

- The light can change direction. It refracts.
- The light can bounce off of the surface. It reflects.
- The light can get stuck in the material. It is absorbed.

Light can hit transparent (trans-PAIR-uhnt) material. Transparent means that you can see through it. This is because light moves through it. This bend in the light is called refraction (ruh-FRAK-shuhn).

Light rays can hit opaque (oh-PAYK) material. That means you can not see through it. Some of the light bounces off. It goes in a new direction. This is called reflection (ruh-FLEK-shuhn). Smooth surfaces like mirrors reflect more light. Surfaces that are rough reflect less light.

When light reflects from a very smooth surface, all of the light rays reflect in the same direction. A mirror is smooth, so you can see yourself in it.

When light reflects from a rough surface, the rays reflect in many directions. This is why you can not see your reflection in paper. The surface is too rough.

Color

Color is all about absorption (uhb-ZORP-shuhn). Look at the clothes you're wearing. What colors are they? You may think the color is in the clothing, but that is not how it works. The colors come from light. The light reflects off the clothes and comes to our eyes. We see the colors of the light.

You know that light is made of waves. Each color has its own frequency. White light has light of all the colors' frequencies. When white light strikes an object, each frequency does its own thing. Some frequencies get absorbed. They are not seen. Some reflect and travel to our eye. We only see the ones that get to our eyes.

White light is made of all the colors of the rainbow. These colors are red, orange, yellow, green, blue, indigo, and violet. Some people know the colors as ROYGBIV. Now, look at something red. You can see that it absorbs the frequencies for OYGBIV. Only R, or red, is reflected. The red light comes back to your eye, and you see red.

The important idea is that the color is not in the object. It is in the reflected light.

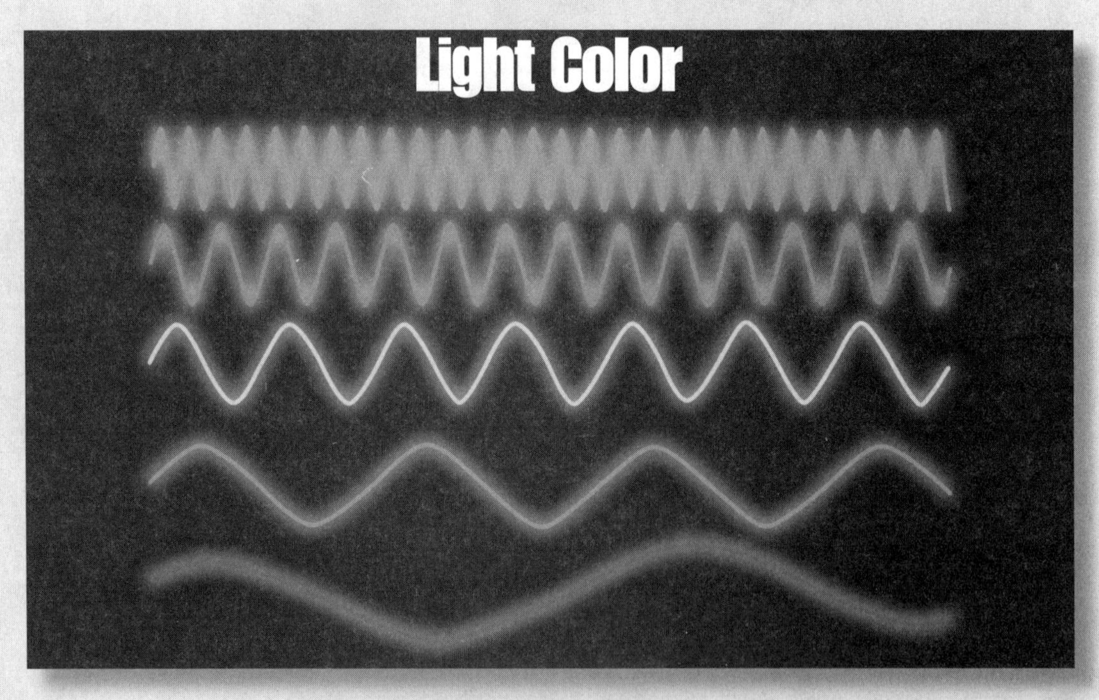

frequencies and colors of light

Comprehension Question

Describe the three things that light does when it strikes an object.

Radiant Light

The energy of light is called radiant energy. To radiate means to send out rays or waves. Only a certain type of radiant energy can be seen with the human eye. We call this visible light. Visible means that we can see it.

We can see because of light. Light waves bounce off objects and travel to our eyes. Our eyes and brain work together to translate that light into what we see.

Light travels in waves much like water moves in waves. The amount of energy that a wave carries determines the color of the light. Waves differ from each other in length, rate, and size. These are called wavelength, frequency (FREE-kwuhn-see), and amplitude (AM-pli-tood). Frequency relates to the color of the light.

Refract, Reflect, Absorb

What happens when a light wave hits the atoms that make up everything? Several things might happen:

- The light can change direction, or refract.
- Some of the light rays can reflect off of the surface.
- The light can be absorbed into the material.

Light rays bend as they travel through the surface of transparent (trans-PAIR-uhnt) material. Transparent means that light can be seen through it and move through it. This bend in the light is called refraction (ruh-FRAK-shuhn). It occurs when light travels through different materials at different speeds.

The return of a wave of energy after it strikes a surface is called reflection (ruh-FLEK-shuhn). Smooth and polished surfaces like mirrors reflect more light than surfaces that are rough or bumpy.

When light reflects from a smooth surface, all of the light rays reflect in the same direction. A mirror is smooth, so you can see your image in it. When light reflects from a rough surface, the rays reflect in many directions. It is impossible to see your reflection in paper, because the surface is rough.

Color

When it comes to color, absorption (uhb-ZORP-shuhn) is the key. Look at the clothes you're wearing. What colors are they? The truth is, the colors are not in the clothing. The colors come from reflected and absorbed light. We see the colors because of the light that is reflected and sent to our eyes.

You know that light is made of waves. Each color has its own frequency. When visible light strikes an object, each frequency behaves differently. Some frequencies are absorbed. They are not seen. Some are reflected. The reflections are what appear as the color or colors of an object.

White light is made of all the colors of the rainbow. These colors are red, orange, yellow, green, blue, indigo, and violet. Some people know the colors as ROYGBIV. Now, look at something red. You can see just by looking at it that the object absorbs the frequencies for OYGBIV. Only R, or red, is reflected. Your eyes pick up that reflection, and you see the object as red.

The important idea is that the color is not in the object. It is in the reflected light.

Describe how you can see light reflected, refracted, and absorbed.

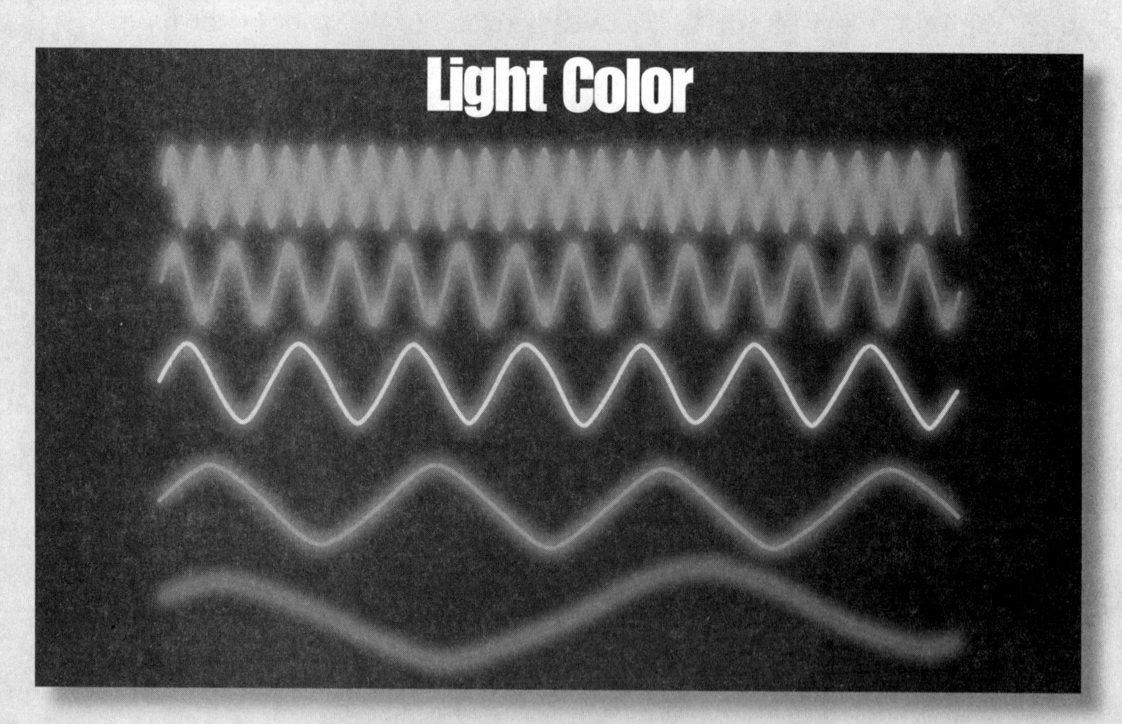

frequencies and colors of light

Comprehension Question

Describe how you can see light reflected, refracted, and absorbed.

Radiant Light

The energy of light is radiant energy, which means it sends out rays or waves. Only a certain type of radiant energy, called visible light, can be seen with the human eye.

The human eye perceives light waves; we can see because those light waves bounce off objects before arriving at our eyes. Our eyes and brain work together to translate the light we detect into the objects we perceive.

Light travels in waves somewhat like water moves in waves. Just as there are different ocean waves, there are different light waves. Waves' different characteristics include length, called wavelength; rate, called frequency (FREE-kwuhn-see); and size, called amplitude (AM-pli-tood). The frequency of the light determines the color.

Refract, Reflect, Absorb

What happens when a light wave hits the atoms that make up everything? The energy in the wave cannot be destroyed. It must go somewhere or do something. Several things might happen. More than one may happen at the same time!

- The light can change direction, or refract.
- Some of the light rays can reflect off of the surface.
- The light can be absorbed into the material.

Light rays bend when they travel through transparent (trans-PAIR-uhnt) material. This bend in the light is called refraction (ruh-FRAK-shuhn); it happens because light travels through different materials at different speeds.

The return of a wave of energy after it strikes a surface is reflection (ruh-FLEK-shuhn). Smooth and polished surfaces like mirrors reflect more light than surfaces that are rough or bumpy. When light reflects from a very smooth surface, all of the light rays reflect in the same direction. A mirror is so smooth, it reflects the light perfectly so you can see yourself. When light reflects from a rough surface, the rays reflect in many directions. It is impossible to see your reflection in paper because the surface is rough.

Color

Absorption (uhb-ZORP-shuhn) is the DNA of color. Consider what colors are in the clothing you're wearing. The truth is, the colors are not in the clothing; the colors come from reflected and absorbed light. We see the colors because of the light that is reflected–and the light that isn't reflected–to our eyes.

Each color of light has its own frequency, and when visible light strikes an object, each frequency behaves differently. Some frequencies are absorbed and are never seen again; some are reflected, and these are the colors that we see.

White light is made of all the colors of the rainbow. These colors are red, orange, yellow, green, blue, indigo, and violet. Some people know the colors as ROYGBIV. A red object absorbs all the frequencies for OYGBIV. Only R, or red, is reflected; your eyes pick up that reflection, and you see the object as red. The important idea is that the color is not in the object; it is in the reflected light.

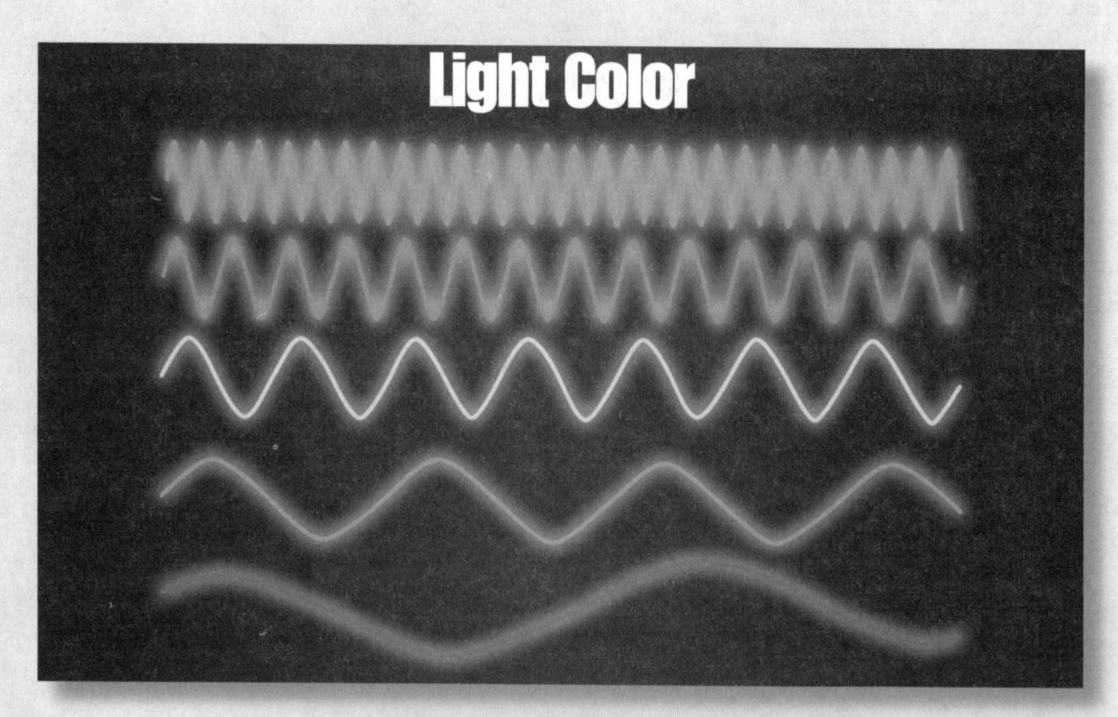

frequencies and colors of light

Comprehension Question

How are you using reflection, refraction, and absorption right now to read this question?

Gravity

When you walk around, you are "stuck" to the ground. You can jump up. You always come back down. Why is this?

Isaac Newton was a scientist. He saw that planets go around the sun. He saw how things fall to Earth. He saw that they were the same. They happened because of the same force. It is called gravity (GRAV-ih-tee).

There is a story about Newton and gravity. He saw an apple fall from a tree. He saw that the apple and the moon are alike. Gravity pulls them both to Earth.

Gravity is what holds us on the ground. It also keeps Earth going around the sun. It keeps the moon going around Earth. Without gravity, things would just float in space. Day to day life would be hard.

Big Gravity

Newton watched the planets in the sky. He saw that the planets moved. They were pulled toward the sun. The planets closest to the sun were pulled the strongest.

He also watched the moon. The moon moves around and around the Earth. But why does its path curve around? Why doesn't it move in a straight line? Newton's first law of motion said it should. Everything moves in a straight line. Things turn only when other forces make them turn. What was the other force?

Newton had an idea about what must be changing the moon's direction. It is the Earth's gravity. Earth's gravity gives the moon a tug. It pulls the moon around in a circle. Think of a ball on a string. You can swing the ball around your head. The string acts like gravity. It keeps the ball moving in a circle. You could let go of the string. The ball flies off at an angle. What would happen to the moon if Earth's gravity stopped working?

Earth's gravity doesn't make the moon fall down. The moon doesn't crash into Earth. The moon has energy. The energy is in its motion. If the moon stopped moving, it would have no energy. It would fall into Earth. Think of the ball again. If you let the ball slow down, it will fall toward you!

Little Gravity

Everything in the universe has gravity. Everything pulls on everything else. How strong that pull is comes from two things. It comes from the thing's mass. It comes from how far away the thing is. A big thing has a big pull of gravity. A faraway thing has a smaller pull of gravity.

The biggest, closest thing to you right now is the Earth. Because it's big and it's close, you feel its pull the strongest.

You have gravity, too. You can't feel it, but it is there. You have mass, so you have gravity. Your gravity is very small. It is hard to see close to Earth. Out in space, you could have your own moon that orbited around you. Maybe it could be your little sister!

Comprehension Question

What is gravity?

Gravity

When you walk around, you are "stuck" to the ground. You can jump, but you always land on the ground. Why is this?

Isaac Newton was the first to understand that the force that makes planets go around the sun is the same force that makes things fall to Earth. It is called gravity (GRAV-ih-tee). There is a story about how Newton came up with the idea of gravity. He saw an apple fall from a tree. He saw that the apple and the moon are similar. Gravity attracts them both to Earth.

Gravity is what holds us on the ground. It keeps us from floating into space. It also keeps Earth going around the sun and the moon going around Earth. Without gravity, things would just bob in space. Everyday life would be difficult.

Big Gravity

Newton watched the planets in the sky. He saw that the planets were pulled toward the sun. The closer the planets were to the sun, the stronger the pull on them.

He also watched the moon. He knew that the moon was moving around and around Earth. But why did its path curve around? Why did it not shoot away from Earth in a straight line? Newton's first law of motion said it should. The moon should move in a straight line unless some other force makes it change direction. What was the other force?

Newton figured out what must be changing the moon's direction: Earth's gravity. Earth's gravity gives the moon a constant tug. It pulls it around in a circle. Think of a ball on a string. You can swing the ball around your head. The pull on the string acts like gravity. It keeps the ball moving in a circle. You could let go of the string, too. Then the ball would fly off at an angle. What would happen to the moon if Earth's gravity stopped working?

Earth's gravity doesn't make the moon fall down and land on Earth, though. This is because the moon has energy from its motion. If the moon stopped moving, then it would start falling toward Earth. Think of the ball again. If you let the ball slow down, it will fall towards you!

Little Gravity

Everything in the universe has gravity. Everything pulls on everything else. The strength of that pull comes from two things. It comes from the object's mass and how far away it is. A big object has a big pull of gravity. A faraway object has a smaller pull of gravity. The biggest, closest thing to you right now is Earth. Because it's big and it's close, you feel its pull the strongest.

You can't feel it, but you have gravity, too. You have mass, so you have gravity. Compared to the Earth, your gravity is very small. If you were floating out in space, though, you could have your own moon that orbited around you–maybe your little sister!

Comprehension Question

How does gravity keep the moon in place?

Gravity

When you walk around, you are "stuck" to the ground. You can jump, but the ground always gets you in the end. Why is this?

Isaac Newton was the first to realize that the force that makes planets go around the sun is the same force that makes things fall to Earth. It is called gravity (GRAV-i-tee). A story about Newton says that he figured out gravity when he saw an apple fall from a tree. He realized that the apple and the moon are similar. Gravity attracts them both to Earth.

Gravity is what holds us on the ground and keeps us from floating into space. It also keeps Earth going around the sun and the moon going around Earth. Without gravity, things would just bob in space. Everyday life would be difficult.

Big Gravity

Newton watched the planets in the sky. He figured out that the planets were pulled toward the sun. The closer the planets were to the sun, the stronger the pull on them.

He also watched the moon. He knew that the moon was moving around and around Earth. But why did its path curve around? Why did it not shoot away from Earth in a straight line? That was Newton's first law of motion: objects move in straight lines unless some other force makes them change direction.

Newton figured out what must be changing the moon's direction: Earth's gravity. Earth's gravity gives it a constant tug, which pulls it around in a circle. Think of a ball on a string. If you swing the ball around your head, then the tension in the string acts like gravity. It keeps the ball moving in a circle. If you were to let go of the string, then the ball would fly off at an angle. This is what would happen to the moon if Earth's gravity stopped working.

The reason Earth's gravity doesn't make the moon fall down and land on Earth is that the moon has energy from its motion. If something came along that stopped the moon from moving, then it would start falling toward Earth. Just as in the ball example above, letting the ball slow down too much makes it collapse toward you!

Little Gravity

Everything in the universe exerts a pull of gravity on everything else. The strength of that pull is determined by the object's mass and how far away it is. A big object has a big pull of gravity. A faraway object has a smaller pull of gravity. The biggest, closest thing to you right now is Earth. Because it's big and it's close, you feel its pull the strongest.

You can't feel it, but you have gravity, too. You have mass, so you have gravity. Compared to Earth, though, your gravity is very small. If you were floating out in space, though, you could have your own moon that orbited around you–maybe your little sister!

Comprehension Question

Describe the forces that keep the moon in its orbit.

Gravity

Have you ever wondered why you are "stuck" to the ground? You can jump, but you always land; you can pilot an airplane, but it can't fly forever. The ground always gets you back in the end; why?

Isaac Newton was the first to realize that the force that makes planets go around the sun is the same force that makes things fall. It is called gravity (GRAV-i-tee). Newton first comprehended gravity when he observed an apple fall from a tree. He realized that the apple and the moon are similar because gravity attracts them both to Earth.

Gravity is what anchors us to the ground and prevents us from floating into space. Gravity keeps the moon orbiting around Earth and Earth itself orbiting around the sun. Without gravity, things would just bob listlessly in space—making everyday life difficult!

Big Gravity

Newton watched the planets in the sky. He figured out that the planets were attracted by the sun and were compelled to move toward it. The closer the planets were to the sun, the stronger the attraction.

He also watched the moon orbit around Earth. Why did its path curve around, he wondered. Why did it not shoot away from Earth in a straight line? That was Newton's first law of motion: objects move in straight lines unless some other force intervenes and makes them change direction.

Newton determined what must be changing the moon's direction: Earth's gravity. Earth's gravity exerts a constant attraction on the moon, which pulls it around in a circle. Imagine a ball on a string, swung around your head: the tension in the string acts like gravity. It constrains the ball's motion and keeps it moving in a circle. Releasing the string would send the ball careening off at an angle. This is identical to what would happen to the moon if Earth's gravity mysteriously stopped working.

The reason Earth's gravity doesn't pull the moon down to the surface of Earth is that the moon has kinetic energy from its motion. If some extraterrestrial force stopped it from moving, only then would the moon start falling toward Earth. Consider the ball example again: if you let the ball slow down too much, then its orbit will collapse toward you!

Little Gravity

Everything in the universe exerts a gravitational pull on everything else. The strength of that pull is determined by the inverse proportion of the objects' mass and their distance apart. Massive objects have big gravitational force. Distant objects have smaller gravitational power. The most massive, closest thing to you right now is the planet Earth. Because it's massive and it's close, it exerts the strongest force on you.

You can't feel it, but you have your own gravitational pull: you have mass, so you have gravity. Compared to the Earth, however, your gravity is very small. If you were floating out in space, you could have your own moon that orbited around you–maybe your little sister!

Comprehension Question

Describe how gravity works.

Relativity

No Ether!

In 1905, Albert Einstein wrote a paper. It explained his special theory of relativity (rel-ah-TIV-uh-tee). This paper was about how people thought of light.

People thought that light waves worked like ocean waves. Ocean waves move through the water. They thought light waves had to move through something, too. They called it ether. There was one problem. Two scientists tried to measure the ether. They couldn't find any. There wasn't any. It didn't exist.

Einstein's paper started with the work of others. He used math. He showed that light did not need to move through things. It did not move like ocean waves. It always went the same speed. That speed was a very big number. Instead of always writing the number, Einstein just wrote c.

Spacetime

Einstein wasn't done, though. Light was not like ocean waves at all!

If you were on a boat, you could move along with the waves. If you stood on the deck, the waves might look like they did not move. The boat could even move faster than the waves. Then, the waves would look like they were going backward!

Light does not work that way. Light always looks like it moves at c. Imagine you have a fast bicycle and your friend has a flashlight. Your friend shines the flashlight and you pedal after the beam of light. It doesn't matter how fast you go. The light will still shine ahead as if you were standing still. How could that be? Einstein had the answer.

Your bicycle cannot go as fast as light.

His answer sounds crazy. He said the faster you go, the slower time goes. And everything gets shorter! The front of your bike will shrink. It will smash back toward the end of your bike. The more you try to catch the light, the slower time goes and the shorter you get. We don't notice this in everyday life. This only happens close to the speed of light.

Einstein showed that the old ideas about space and time had to be changed. They are two parts of the same thing, called spacetime. The faster you go, the weirder things get!

$E=mc^2$

Einstein wrote one more paper. The next one showed how energy and mass were linked. It says that something's energy (E) is equal to its mass (m) times the speed of light (c) squared. This is written $E = mc^2$.

The paper showed that mass and energy are linked. It means that energy acts the same way as mass. Not only were space and time the same thing. Mass and energy were the same thing, too!

You can see this in a solar eclipse. This occurs when the moon comes between the sun and Earth. It blocks the sun's light. You can see that starlight bends when it passes the sun. Light is energy. It falls toward the sun, just like mass does.

The faster you go, the weirder things get!

Comprehension Question

What were Einstein's two big ideas?

Relativity

No Ether!

In 1905, Albert Einstein wrote a paper that explained his special theory of relativity (rel-ah-TIV-uh-tee). It solved a problem with how scientists understood light.

People thought that light waves worked like ocean waves. Ocean waves move through the water. They thought light waves had to move through something, too. They called it ether. There was one problem. Two scientists tried to measure the ether. They couldn't find any. There wasn't any. It didn't exist.

Einstein's paper started with the work of others. He used math to show that light did not need to move through things. It was different than ocean waves. It always went the same speed. That speed was a very big number. Instead of always writing the number, Einstein just wrote *c*.

Spacetime

Einstein wasn't done, though. Light was very different than ocean waves!

If you were on a boat, you could move along with the waves. If you stood on the deck, the waves might look like they did not move. The boat could even move faster than the waves. Then, the waves would look like they were going backward!

Light does not work that way. Light always looks like it moves at *c*. Imagine you have a fast bicycle and your friend has a flashlight. Your friend shines the flashlight and you pedal in the same direction. It doesn't matter how fast you go. The light will still shine ahead as if you were standing still. How could that be? Einstein had the answer.

Your bicycle cannot go as fast as light.

His answer sounds crazy. He said the faster you go, the slower time goes. And everything gets shorter! The front of your bike will shrink back toward the end of your bike. The more you try to catch the light, the slower time goes and the shorter you get. We don't notice this in everyday life. This only happens close to the speed of light.

Einstein showed that the old definitions of space and time had to be changed. They are two parts of the same thing, called spacetime. The faster you go, the weirder things get!

$E=mc^2$

Einstein wrote two papers that year. The second one explained the relationship between energy and mass. It says that something's energy (E) is equal to its mass (m) times the speed of light (c) squared. This is written $E = mc^2$.

The paper showed that mass and energy are linked. This means that energy acts the same way as mass. Not only were space and time the same thing. Mass and energy were the same thing, too!

You can see this in a solar eclipse. This occurs when the moon comes between the sun and Earth. It blocks the sun's light. You can see that starlight bends when it passes the sun. The light energy falls toward the sun just like mass does.

The faster you go, the weirder things get!

Comprehension Question

What did Einstein's two papers explain?

Relativity

No Ether!

In 1905, Albert Einstein published a paper that explained his special theory of relativity (rel-uh-TIV-uh-tee). It solved a problem with how people understood light.

At the time, scientists believed that light waves worked like waves in the ocean. Ocean waves travel through the water. They thought light waves had to travel through something, too. They called this substance ether. There was one problem. Two scientists made an experiment to measure the ether. They couldn't find any.

Einstein's paper started with the work of other scientists. He used their math to explain that light did not need to travel through anything. Light was different from ocean waves. Light always traveled at the same speed. That speed was a very big number. Instead of always writing the number, Einstein just wrote c.

Spacetime

Einstein wasn't done yet: light was very different from ocean waves!

Imagine you were on a sailing ship, traveling alongside the waves. From your perspective on deck, the waves might appear to stop moving. You could even travel faster than the waves; then, the waves would appear to go backward!

Light works differently. Light always looks like it travels at c. Imagine you have a fast bicycle and your friend has a flashlight. Your friend shines the flashlight and you pedal in the same direction. No matter how fast you go, the light will still shine ahead as if you were standing still.

Your bicycle cannot go as fast as light.

Einstein said that the faster you go, the slower time goes and the shorter everything gets. The front of your bike will shrink back toward the end of your bike. The more you try to catch the light, the slower time goes and the shorter you get. We don't notice this in everyday life; it only happens close to the speed of light.

Einstein demonstrated that the old definitions of space and time needed to be changed. Since they are connected, Einstein suggested calling them spacetime.

$E=mc^2$

Einstein published another paper that year to explain the relationship between energy and mass. It says that the energy of a body (E) equals its mass (m) times the speed of light (c) squared. This is written $E = mc^2$.

The equation shows that mass and energy are linked. It means that energy must behave the same as mass does. Not only were space and time the same thing. Mass and energy were the same thing, too!

This can be observed during a solar eclipse. A solar eclipse occurs when the moon comes between the sun and Earth, blocking the sun's bright light. During this time, we can see that starlight bends when it passes the sun. Light is energy that falls toward the sun, just as mass does.

The faster you go, the weirder things get!

Comprehension Question

How were Einstein's two papers important?

Relativity

No Ether!

In 1905, Albert Einstein published a paper that explained his special theory of relativity (rel-uh-TIV-uh-tee). It unraveled a problem with how science understood light.

At the time, scientists believed that light waves worked like waves in the ocean. Since ocean waves travel through the water, they thought light waves had to travel through something, too. They called this substance ether, but there was one problem. Two scientists made an experiment to measure the ether, and they couldn't find any.

Einstein's paper started with the work of other scientists. He used their ideas and some mathematics to explain that light did not need to travel through anything. Light was different from ocean waves: it always traveled at the same speed. That speed was a very big number, so instead of always writing the number, Einstein just wrote c.

Spacetime

Einstein wasn't done yet; light was very different from ocean waves!

Imagine you were on a sailing ship, traveling alongside the waves. From your perspective on deck, the waves might appear to stop moving. You could even travel faster than the waves; then the waves would appear to go backward!

Light works differently. Light always appears to travel at c. Imagine you have a fast bicycle and your friend has a flashlight. Your friend shines the flashlight and you pedal in the same direction. No matter how fast you go, the light will still shine ahead as if you were standing still.

Your bicycle cannot go as fast as light.

When you approach the speed of light, things start getting really weird. Einstein said that the faster you go, the slower time goes and the shorter everything gets. The front of your bike will shrink back toward the end of your bike. The more you try to catch the light, the slower time goes and the shorter you get. We don't notice this in everyday life; it only happens close to the speed of light.

Einstein demonstrated that the old definitions of space and time needed to be changed. Since they are connected, Einstein suggested calling them spacetime.

$E=mc_2$

Einstein published another paper that year to explain the relationship between energy and mass. It says that the energy of a body (E) equals its mass (m) times the speed of light (c) squared. This is written $E = mc^2$.

The equation shows that mass and energy are linked. It means that energy must behave the same as mass does. Not only were space and time the same thing. Mass and energy were the same thing, too!

This can be observed during a solar eclipse. A solar eclipse occurs when the moon comes between the sun and Earth, blocking the sun's bright light. During this time, we can see that starlight bends when it passes the sun. Light is energy that falls toward the sun, just as mass does.

The faster you go, the weirder things get!

Comprehension Question

How were Einstein's two papers similar?

Electromagnetism

Electricity

Electricity is made by electrons. Electrons flow from one place to another place. They are called a current when they flow. They flow in a circuit (SIR-kit). A circuit is a closed loop. It is made of stuff that can move the current.

There are two types of electricity. One type is static electricity. One type is current electricity. Do this: shuffle your feet across a carpet. Then touch your friend's hand. You may both feel a shock. This shock is static electricity.

Electricity either rests or moves. It is called static when it rests. It is at rest until you move your feet on the carpet. Then the electrons move. They move from one thing to another thing. At first, one thing has a positive charge. One has a negative charge. The shock you feel when you touch hands is made by electrons. They move from one hand to the other. This evens out the charge. Now both have no charge.

Think about a river that runs in a circle. Current electricity is like the river. The electrons are on the move. We can study magnets to see how that works.

Magnetism

Have you used a magnet? What does it do? Magnets make energy that you can not see. The force only affects some things. Iron is one of these things. The force can move a piece of iron. But nothing has to touch the iron to do it.

The pull of a magnet can only go so far. The pull is called its magnetic field. The force of a magnet is only felt in the field.

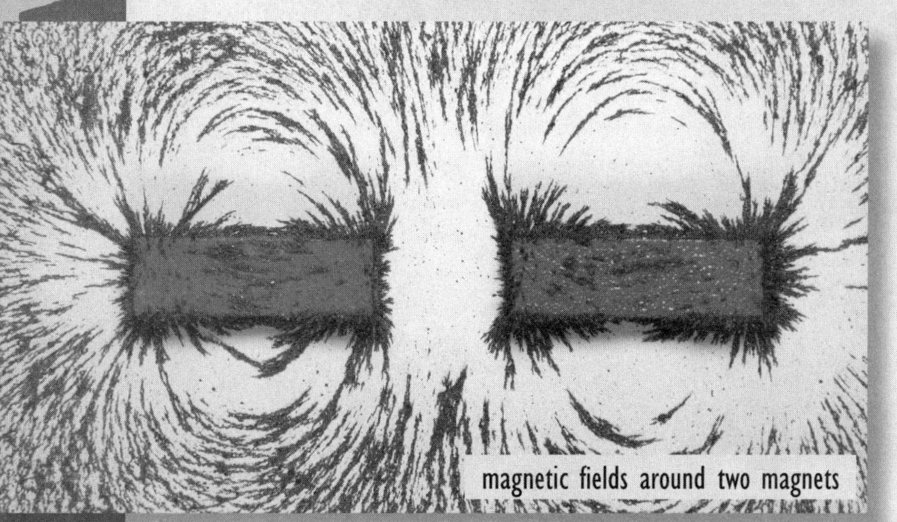

magnetic fields around two magnets

The first magnets were found in nature. No one made them. Then people tried to make fields like magnets do. One man found a way. Hans Oersted (UR-sted) put a compass near a current. The magnet in the compass moved. Oersted wanted to know more about it. He learned that currents have magnetic fields.

Electromagnetism

This was big news. It showed the link between electricity and magnetism. It also led to something new. It led to the electromagnet.

An electromagnet can be a simple thing. It just needs two things to be made. It needs a coil of wire and a battery. They are hooked to each other. Electricity flows through the wire. It makes a magnetic field. The magnet can be made stronger. More turns in the coil will do it. More current in the circuit will also do it. A nail through the coil can make it even stronger. Magnets are found in many things like phones and washing machines.

an early electromagnet

Comprehension Question

What is an electromagnet?

Electromagnetism

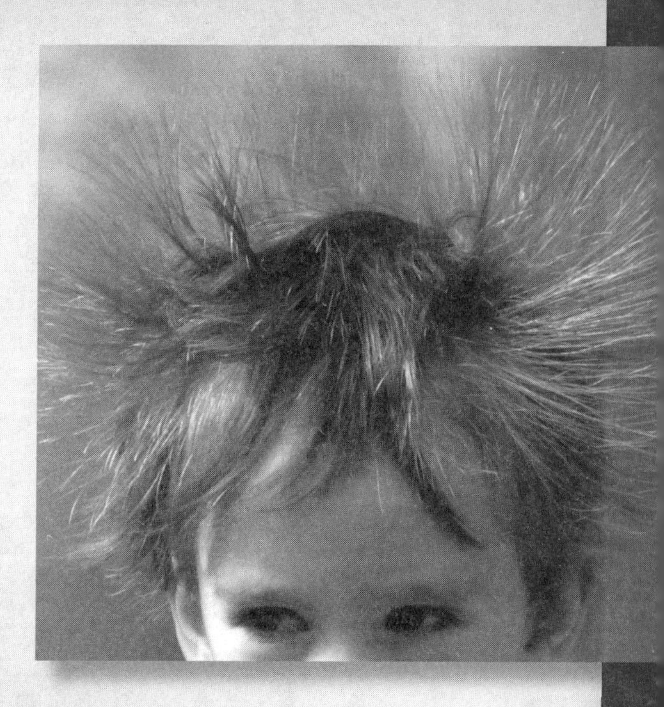

Electricity

An electrical current is made by the flow of electrons. They move from one place to another. There must be an electrical circuit (SIR-kit) for a current to flow. A circuit is a closed loop of material. It moves the flow.

There are two types of electricity. One type is static electricity. One type is current electricity. Do this: shuffle your feet across a carpet. Then touch your friend's hand. You may both feel a shock. This shock is a jolt of static electricity.

Electricity is at rest until it is able to move. It is called static when it is at rest. It is at rest before you move your feet on the carpet. You move electrons from one surface to the other when you shuffle your feet. Each surface has a different charge. One surface has a positive charge. The other surface has a negative charge. This difference in charges is called a "potential difference." The jolt you feel when you touch your friend's hand is made by electrons. They move from one hand to the other. This evens out the difference. It makes both surfaces neutral again. That means they have no charge.

Current electricity is like a river that runs in a circle. The electrons are on the move. We can study magnets to see how that works.

Magnetism

Have you ever played with magnets? Magnets create a force that you cannot see. The force only affects some things. Iron is one of these things. Magnetic forces can move a piece of iron. Nothing has to touch the metal to do it.

The reach of a magnet can only go so far, though. A magnet's reach is called its magnetic field. Magnetic forces can only be felt within the field.

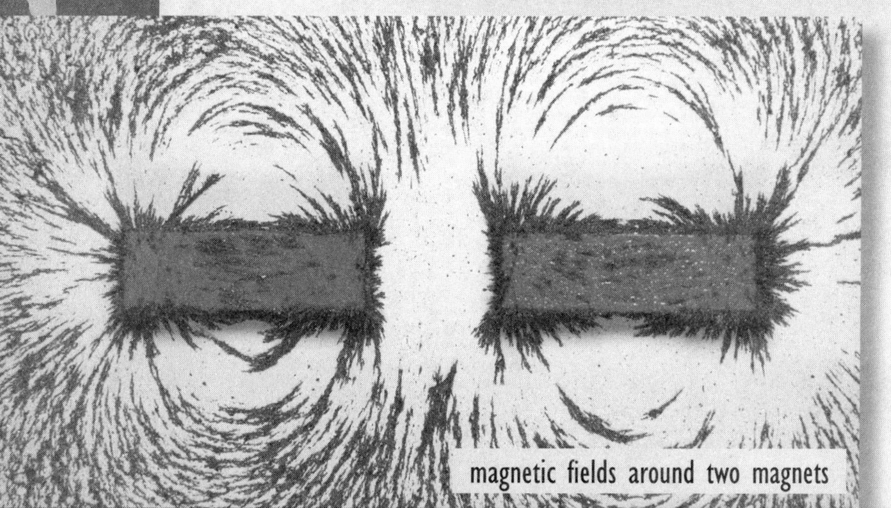
magnetic fields around two magnets

The first magnets were found in nature. People began to wonder if they could make magnets themselves. One scientist found a way. Hans Oersted (UR-sted) put a compass near a current. The magnet in the compass moved. The current had made a magnetic field. Oersted found that electrical currents have magnetic fields.

Electromagnetism

Oersted's work was big news. It showed that electricity and magnetism have ties to each other. It also led to a new find. That new find was the electromagnet.

An electromagnet can be simple to make. It is a coil of wire attached to a battery. A current flows through the wire. It makes a magnetic field. An electromagnet can also be made stronger. More turns in the coil will make it stronger. More current in the circuit will also do it. A nail through the coil can make it stronger still. Electromagnets are used in many things. Phones and washing machines are just two of those things.

an early electromagnet

Comprehension Question

How does an electromagnet combine electricity and magnetism?

Electromagnetism

Electricity

An electrical current is the flow of electrons from one place to another. There must be an electrical circuit for a current to flow. A circuit is a closed loop of conducting material. Electricity can flow along it. There are two types of electricity. They are static and current.

Shuffle your feet across a carpet. Then touch your friend's hand. You may both feel a small shock. This shock is a tiny jolt of static electricity.

Electricity is at rest until it is able to move. It is called static when it is at rest. You move electrons from one surface to the other when you shuffle your feet on the carpet. This makes one surface have a positive charge and the other have a negative charge. This difference in charges is called a "potential difference." When you touch your friend's hand, the jolt you feel is the electrons moving from one hand to the other. This evens out the potential difference. It makes both surfaces neutral again.

Current electricity is like a river that runs in a circle. The electrons are always moving. To see how that works, we need to understand magnets.

Magnetism

Have you ever played with magnets? Magnets create an invisible force. The force only affects some things. Iron is one of these things. Magnetic forces can move a piece of iron without anything touching the metal.

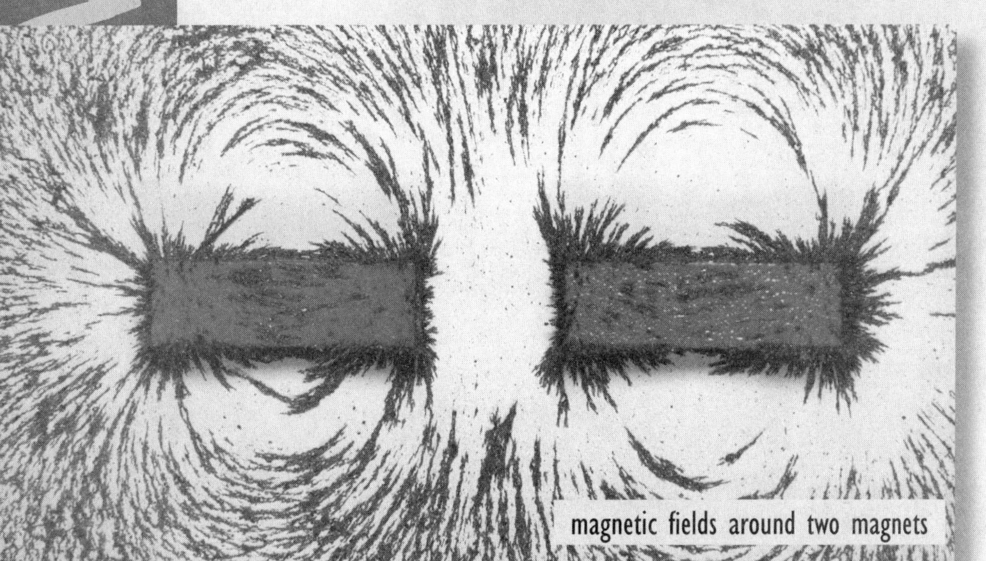
magnetic fields around two magnets

Magnetism can only reach so far, though. The reach of a magnet is called its magnetic field. Magnetic forces can only be felt within the field. The lines of force cannot be seen.

The first magnets were found in nature. Scientists began to wonder if they could make artificial ones. Artificial means something that is made by people.

One scientist found a way. In 1820, Hans Oersted placed a compass near an electrical current. He saw that the needle on the compass moved. The current had made a magnetic field. Oersted studied this some more. He found that electrical currents have magnetic fields.

Electromagnetism

This showed that there is a close link between electricity and magnetism. It also led to a new discovery. That was the electromagnet.

A simple electromagnet is a coil of wire that is attached to a battery. Electricity flows through the wire. It makes a magnetic field. An electromagnet can be made stronger. One way to make it stronger is to use more turns in the coil. Another way is to add more current in the circuit. A piece of soft iron like a nail through the coil makes the electromagnet stronger still. Electromagnets are used in many things. Telephones and washing machines are just two of those things.

an early electromagnet

Comprehension Question

Describe electromagnetism.

Electromagnetism

Electricity

An electrical current is the flow of electrons from one place to another. For a current to flow, there must be an electrical circuit. This is a closed loop of conducting material that the electricity can flow along. There are two types of electricity: static and current.

Shuffle your feet across a carpet. Then touch your friend's hand. You may both feel a small shock. This shock is really a tiny jolt of static electricity.

Until electricity is able to move, it is at rest. That is called static. When you shuffle your feet on a carpet, you transfer electrons from one surface to the other. This makes one surface positively charged and the other negatively charged. This difference in charges is called a "potential difference." When you touch your friend's hand, the jolt you feel is the electrons moving from one hand to the other. This evens out the potential difference and makes both surfaces neutral again.

Current electricity is like a river that runs in a circle. The electrons are always moving. To see how that works, we need to understand magnets.

Magnetism

You've probably played with magnets before. Magnets create an invisible force that only affects certain things. Iron is one of these things. Magnetic forces can move a piece of iron without anything touching the metal.

Magnetism can only reach so far, though. The reach of a magnet is called its magnetic field. Magnetic forces can be felt within the field but not outside it. A magnetic field is made of invisible lines of force.

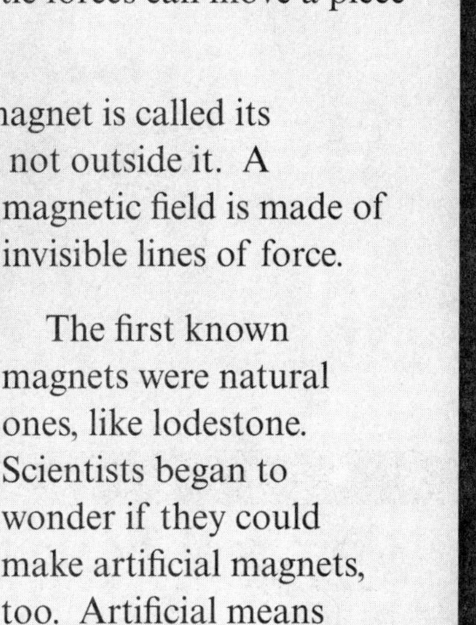

The first known magnets were natural ones, like lodestone. Scientists began to wonder if they could make artificial magnets, too. Artificial means something that is man-made.

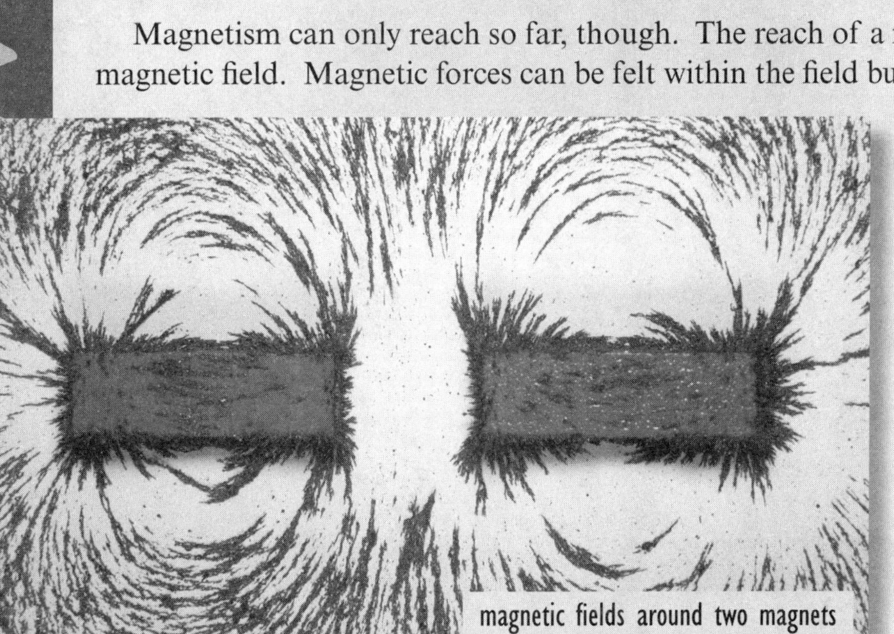

magnetic fields around two magnets

In 1820, one scientist found a way. At a party, Hans Oersted placed a compass near an electrical current. He noticed that the needle on the compass moved. The electrical current had made a magnetic field. Oersted investigated further. He found that electrical currents have magnetic fields that go around the wire.

Electromagnetism

This showed that there is a close relationship between electricity and magnetism. And that led to the discovery of the electromagnet.

A simple electromagnet is a coil of wire attached to a battery. The flow of electricity through the coils of wire creates a magnetic field. An electromagnet can be made stronger by using more turns in the coil or more current in the circuit. A piece of soft iron like a nail put through the coil makes the electromagnet stronger still. Today, electromagnets are found in everything from telephones to the motors in washing machines.

an early electromagnet

Comprehension Question

How are electricity and magnetic fields related? Give examples.

Newton's Laws of Motion

Isaac Newton is famous for three laws. They are about the way things move. He didn't write the laws. Other people called them Newton's Laws of Motion.

Newton's First Law

The first law is about inertia (in-UR-shuh). Inertia means not changing. The law says that a thing will keep doing what is doing. It will only change when something else makes it change. This works for things that are not moving. It works for things that are moving. Things that are not moving don't start moving. Things that are moving keep moving. They move in a straight line with the same speed.

Newton's Second Law

The second law is about acceleration (AK-sell-er-ay-shun). It says what happens when something is pushed. It does one of two things. The force may make the thing speed up. It may make it slow down. A bigger force makes a bigger change. A big force is needed to move something heavy. A small force can move something light the same distance. This makes sense. A bowling ball is harder to throw than a tennis ball. It is harder to stop a car than a bike. The law says one more thing. What gets pushed always moves in the same way as the force is moving.

Newton's Third Law

The third law is about two things. It is about action. It is also about reaction. It says that for every action, there is a reaction. A force pushes on an object. The object always pushes back. This is called the reaction force. The reaction force is always equal. It is always opposite. The object pushes back just as hard. It pushes back in the opposite direction.

This law explains many things. It explains why you can move a rowboat in water. You use an oar. The water pushes back on the oar. It pushes on the oar as much as the oar pushes on the water. This moves the boat. It also tells why a chair stays where it is. It doesn't fall through the floor. The floor pushes up. The floor keeps the chair on top. You can hit a ball with a bat. The ball pushes on the bat. It pushes as much as the bat pushed on it. Hit it just right, and all that force turns into a home run!

Comprehension Question

What are Newton's three laws about?

Newton's Laws of Motion

Isaac Newton is famous for three laws about the way things move. Newton didn't write the laws. Other scientists studying Newton's work wrote them. They called them Newton's Laws of Motion.

Newton's First Law

Newton's First Law of Motion is about inertia (in-UR-shuh). Inertia means not changing movement. The law says that objects keep doing what they are already doing. They only change when something else makes them change. This works for things that are still and for things that are moving. Things that are still stay still. Things that are moving keep moving. They move in a straight line with the same speed.

Newton's Second Law

Newton's Second Law is about acceleration (AK-sell-er-ay-shun). It says what happens when a force pushes an object. The force makes the object speed up or slow down. The bigger the force, the bigger the change. The law says one more thing. The object always moves in the direction of the force. A bigger force is needed to affect a heavier object. A smaller force can move a lighter object the same distance. This makes sense. A bowling ball is harder to throw than a tennis ball. It is harder to stop a car than a bicycle.

Newton's Third Law

Newton's Third Law of Motion is about action and reaction. It says that for every action, there is a reaction. When a force pushes on an object, the object pushes back. This is called the reaction force. The reaction force is always equal and opposite. The object pushes back just as hard. It pushes back in the opposite direction.

This law explains many things. It explains why you can move a rowboat in water with an oar. The water pushes back on the oar. It pushes as much as the oar pushes on the water. This moves the boat. It also explains why a chair stays where it is instead of falling through the floor. The floor pushes back and keeps it there. You can hit a baseball with a bat. The ball pushes on the bat as much as the bat pushes on the ball. Hit it just right, and all that force creates a home run!

Comprehension Question

Give an everyday example of each of Newton's Laws of Motion.

Newton's Laws of Motion

Newton is perhaps most famous for three laws about the way things move. Newton didn't write the laws. Other scientists studying Newton's work wrote them and called them Newton's Laws of Motion. Newtonian Mechanics are based on the Laws of Motion.

Newton's First Law

Newton's First Law of Motion is the law of inertia (in-UR-shuh). Inertia means resistance to changes in motion. The law says that, so long as an unbalanced force doesn't act on an object, then it will keep doing what it's already doing. This works for things that are still and for things that are moving. Things that are still will stay still. Things that are moving will keep moving in a straight line with the same speed.

Newton's Second Law

Newton's Second Law is the law of acceleration. It describes what happens when you apply a force to an object. It says that the bigger the force, the more the object speeds up or slows down. It also says that the object will always move in the same direction of the force. A bigger force is needed to make a heavier object speed up or slow down by the same amount as a light object. This makes sense. For example, a bowling ball is harder to throw than a tennis ball. It is harder to stop a car than a bicycle.

Newton's Third Law

Newton's Third Law of Motion is the law of action and reaction. It says that for every action, there is an equal and opposite reaction. This means that whenever a force pushes on an object, the object pushes back in the opposite direction. The force of the object pushing back is called the reaction force.

This law explains many things. For example, it explains why we can move a rowboat in water with an oar. The water pushes back on the oar as much as the oar pushes on the water. This moves the boat. It also explains why a chair stays where it is instead of crashing through the floor. The floor pushes back and keeps it there. Also, when you hit a baseball with a bat, the ball pushes on the bat as much as the bat pushes on the ball. Hit it just right, and all that force creates a home run!

Comprehension Question

How do you use Newton's Laws of Motion?

Newton's Laws of Motion

Newton is perhaps most famous for three laws about objects in motion and the forces that act on them. Newton didn't write the laws as we know them today; other scientists studying Newton's work wrote them and called them Newton's Laws of Motion. Newtonian Mechanics are based on the Laws of Motion.

Newton's First Law

Newton's First Law of Motion concerns inertia (in-UR-shuh). Inertia, the resistance to changes in motion, has a significant impact on objects in motion. The law says that, so long as an unbalanced force doesn't interfere, then an object will keep doing what it's already doing. This applies to objects whether they are still or moving. Objects that are still remain still. Objects that are moving continue moving with a constant velocity.

Newton's Second Law

Newton's Second Law concerns acceleration. It describes the effect of applying a force to an object. It says that the bigger the force, the more the object accelerates, or changes velocity. Additionally, the object will always move in the same direction of the force. A bigger force is needed to make a heavier object accelerate the same amount as a light object. This makes sense. For example, a bowling ball is more difficult to throw than a tennis ball. It is more difficult to stop a car than a bicycle.

Newton's Third Law

Newton's Third Law of Motion is the law of action and reaction. It says that for every action, there is an equal and opposite reaction. This means that whenever a force pushes on an object, the object pushes back in the opposite direction. The force of the object pushing back is called the reaction force.

This law explains why a rowboat moves when oarsmen row. The oarsmen push backwards on the water with an oar. The water pushes back on the oar with an equal and opposite reaction. This moves the boat forward. It also explains why an armchair sits on the floor instead of crashing through it: the floor pushes back and keeps it there. Alternately, when you hit a baseball with an aluminum bat, the ball has another equal and opposite reaction. Hit it just right, and all that force creates a home run!

Comprehension Question

Describe how an automobile uses all three of Newton's Laws of Motion.

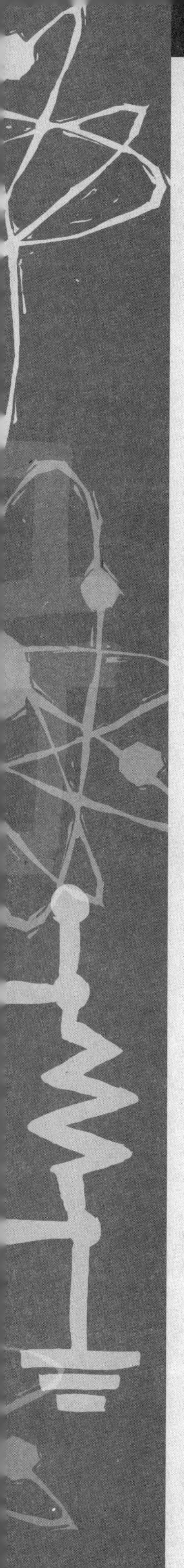

Resources

Works Cited

August, Diane and Timothy Shanahan (Eds). (2006). *Developing literacy in second-language learners: Report of the National Literacy Panel on language-minority children and youth.* Mahwah, NJ: Lawrence Erlbaum Associates, Inc.

Marzano, Robert, Debra Pickering, and Jane Pollock. (2001). *Classroom instruction that works.* Alexandria, VA: Association for Supervision and Curriculum Development.

Tomlinson, Carol Ann. (2000). Leadership for Differentiating Schools and Classrooms, Alexandria, VA: Association for Supervision and Curriculum Development.

Image Sources

Passage	Description	Source	Filename
Atoms	Artist's conception of an atom	Shutterstock (2459231)	artistatom.jpg
Atoms	Oxygen atom	Teacher Created Materials	oxygenatom.jpg
Atoms	Atom diagram	Teacher Created Materials	lithiumatom.jpg
Elements, Molecules, and Mixtures	Iron atom	Teacher Created Materials	ironatom.jpg
Elements, Molecules, and Mixtures	Water molecules	Teacher Created Materials	watermolecules.jpg
Elements, Molecules, and Mixtures	Blood mixture diagram	Teacher Created Materials	bloodmixture.jpg
States of Matter	Water in three states of matter	Teacher Created Materials	statesofmatter.jpg
States of Matter	Solid molecular configuration	Teacher Created Materials	solidconfiguration.jpg
States of Matter	Liquid molecular configuration	Teacher Created Materials	liquidconfiguration.jpg
States of Matter	Gaseous molecular configuration	Teacher Created Materials	gasconfiguration.jpg
States of Matter	Absolute zero in Fahrenheit, Celsius, and Kelvin scales	Teacher Created Materials	absolutezero.jpg
The Periodic Table	Helium Atom	Teacher Created Materials	heliumatom.jpg
The Periodic Table	Dmitri Mendeleév	The Granger Collection (0074697)	dmitrimendeleev.jpg
The Periodic Table	Original periodic table	The Granger Collection (0040384)	originalperiodictable.jpg
Chemical Reactions	Antoine and Marie-Anne Lavoisier	The Granger Collection (0026097)	lavoisiers.jpg
Chemical Reactions	Oxygen atom	Teacher Created Materials	oxygenatom.jpg

Resources *(cont.)*

Image Sources *(cont.)*

Passage	Description	Source	Filename
Chemical Reactions	Water molecule diagram	Teacher Created Materials	watermoleculediagram.jpg
Energy	Three kids on a sled	Photos.com (24243787)	sledding.jpg
Energy	Rollercoaster going down	Photos.com (7692577)	rollercoasterdown.jpg
Energy	Rollercoaster going up	Photos.com (26815483)	rollercoasterup.jpg
Heat	A bunsen burner	Photo Researchers (2K8422)	bunsenburner.jpg
Heat	Mushrooms being sauteéd	Shutterstock.com (915436)	sautee.jpg
Heat	The Sun through wheat	Photos.com (19035885)	sunthroughwheat.jpg
Sunlight	The Sun through trees	Shutterstock (21554)	sunthroughtrees.jpg
Sunlight	Sunspot	NASA	sunspot.jpg
Sunlight	Photon	Teacher Created Materials	photon.jpg
Electrical Circuits	Electrical current diagram	Teacher Created Materials	electricalcurrent.jpg
Electrical Circuits	Electrical circuit diagram	Teacher Created Materials	electricalcircuit.jpg
Electrical Circuits	X-ray of a flashlight with battery and bulb shown	Photo Researchers (BE0220)	flashlight.jpg
Vibrations	Larynx	Rick Nease	larynx.jpg
Vibrations	High- and low-frequency sounds diagram	Teacher Created Materials	soundfrequencydiagram.jpg
Vibrations	Bat using sonar	Photo Researchers (SF4799)	batusingsonar.jpg
Radiant Light	Prism	Shutterstock (727777)	prism.jpg
Radiant Light	Glass of water displaying refraction	Shutterstock (2757257)	refractionthroughwater.jpg
Radiant Light	Light frequencies diagram	Teacher Created Materials	lightfrequencydiagram.jpg
Gravity	Girl in parka falling	Shutterstock (1242698)	falling.jpg
Gravity	Isaac Newton	Library of Congress (USZ62-10191)	newton.jpg
Gravity	Ball on string demonstrating orbital forces	Shutterstock (196120)	ballonstring.jpg
Relativity	Albert Einstein	Library of Congress (USZ62-60242)	einstein.jpg
Relativity	Bicyclist riding down street	Teacher Created Materials	bicyclist.jpg
Relativity	Bicycling approaching speed of light	Teacher Created Materials	fastbicyclist.jpg
Electro-magnetism	Boy's hair standing on end due to static electricity	Shutterstock (312815)	staticelectricity.jpg
Electro-magnetism	Two magnets and iron filings showing magnetic fields	Shutterstock (44389)	twomagneticfields.jpg

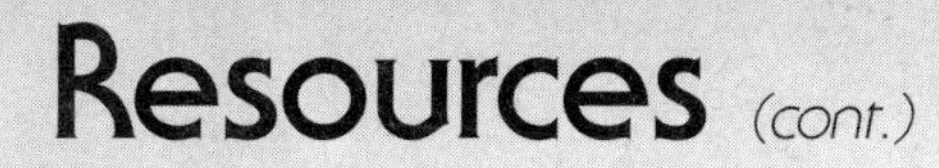

Image Sources (cont.)

Passage	Description	Source	Filename
Electro-magnetism	An old electromagnet	Courtesy of Dr. Greenslade	electromagnet.jpg
Newton's Laws of Motion	Bicyclist moving at a blur	Photos.com (19146001)	blurrybicyclist.jpg
Newton's Laws of Motion	Man on swing	Photos.com (30423331)	manonswing.jpg
Newton's Laws of Motion	Baseball batter	Photos.com (1364655)	baseballbatter.jpg

Contents of Teacher Resource CD

PDF Files

The full-color pdfs provided are each eight pages long and contain all four levels of a reading passage. For example, the Atoms PDF (pages 21–28) is the *atoms.pdf* file.

Text Files

The text files include the text for all four levels of each reading passage. For example, the Atoms text (pages 21–28) is the *atoms.doc* file.

Text Title	Text File	PDF
Atoms	atoms.doc	atoms.pdf
Elements, Molecules, and Mixtures	elements-molecules-mixtures.doc	elements-molecules-mixtures.pdf
States of Matter	states-of-matter.doc	states-of-matter.pdf
Periodic Table	periodic-table.doc	periodic-table.pdf
Chemical Reactions	chemical-reations.doc	chemical-reactions.pdf
Energy	energy.doc	energy.pdf
Heat	heat.doc	heat.pdf
Sunlight	sunlight.doc	sunlight.pdf
Electrical Circuits	electrical-circuits.doc	electrical-circuits.pdf
Vibrations	vibrations.doc	vibrations.pdf
Radiant Light	radiant-light.doc	radiant-light.pdf
Gravity	gravity.doc	gravity.pdf
Relativity	relativity.doc	relativity.pdf
Electromagnetism	electromagnetism.doc	electromagnetism.pdf
Newton's Laws of Motion	newton.doc	newton.pdf

JPEG Files

The images found throughout the book are also provided on the Teacher Resource CD. See pages 141–143 for image descriptions, credits, and filenames.

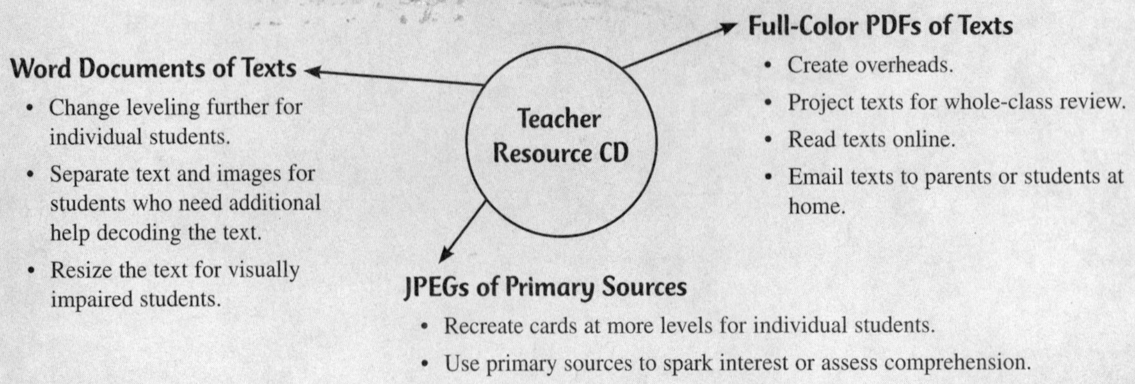

Teacher Resource CD

Word Documents of Texts

- Change leveling further for individual students.
- Separate text and images for students who need additional help decoding the text.
- Resize the text for visually impaired students.

Full-Color PDFs of Texts

- Create overheads.
- Project texts for whole-class review.
- Read texts online.
- Email texts to parents or students at home.

JPEGs of Primary Sources

- Recreate cards at more levels for individual students.
- Use primary sources to spark interest or assess comprehension.